Towards Process Safety 4.0 in the Factory of the Future

*Series Editor*
*Jean-Claude Charpentier*

# Towards Process Safety 4.0 in the Factory of the Future

André Laurent

WILEY

First published 2023 in Great Britain and the United States by ISTE Ltd and John Wiley & Sons, Inc.



ISTE Ltd
27-37 St George's Road
London SW19 4EU
UK

www.iste.co.uk

John Wiley & Sons, Inc.
111 River Street
Hoboken, NJ 07030
USA

www.wiley.com



Library of Congress Control Number: 2022952100

British Library Cataloguing-in-Publication Data
A CIP record for this book is available from the British Library
ISBN 978-1-78630-847-4

# Contents

# Foreword

Experts in the field of safety in process industries have seen their discipline evolve since the mid-2000s. The arrival of Industry 4.0, with its entirely digital control systems, communicating over the Internet, has made it possible to remotely integrate and control production units. This has resulted in significant efficiency and reduced costs; however, it has also opened up inroads for malevolent actors operating anywhere in the world. There is no longer any way of looking at process safety without looking at cybersecurity. Of course, traditional OSH (occupational safety and health) must be also be included in these steps. Thus, industry 4.0 needs a global approach to safety: “Safety 4.0”.

This book, intended for an international audience, attempts to demonstrate the importance of this integration and to define the elements that compose it. It offers examples taken from different points in the highly diversified process industries. These invite the reader to learn general lessons applicable to many other activities. The main point the reader must retain from these examples is the need to involve safety experts from the very beginning of the design of new systems, or to update existing systems. It is not enough to invite them to validate, or worse, add in a “safety” layer to systems designed with no contribution from a safety expert.

Further, and this is also something clearly defined in this book, we will not stop with looking at Industry 4.0 as it exists today in 2022. It is already possible to see examples of the intensification of the process industries and we will undoubtedly discover others, whether this is miniaturization, discontinuous processes being replaced by continuous processes, multi-functionality, or other disruptive approaches and technology. This may pose a considerable challenge to safety – one of the IMPULSE project reports cited in this book declares that “the range of optimal reaction conditions is almost congruent with the danger of a non-controlled reaction”. There is no doubt that some of these developments will contribute directly to safety – the keywords are “intrinsically safe” and “personal safety” – but this book will also show some of the difficulties associated with these keywords.

A final consideration, even with the analytical tools described in this book, is there is a fundamental component of risk management that requires human imagination and the widest possible experience: the identification of dangers. The experience of the European Commission’s Major Accident Hazard Bureau shows us that many accidents in process industries involved a hazard that was not considered during the initial risk analysis.

There is, therefore, quite some work to be done!

Neil MITCHISON
Former Head of the Major Accident Hazard Bureau,
European Commission
Former President, Scientific Council, INERIS

# Preface

The concept of an industry of the future is based on a combination of digital technologies, whose point of intersection is that they make it possible to integrate digital techniques within the functioning of a production unit. This digital integration is an important factor in the development of a new process and for adapting or reconfiguring an existing process as needed. These Industry 4.0 technologies include, among others, equipment and equipment design, as well as systems for acquiring and processing data from the process. The use of these technologies is not risk free. An assessment must be carried out, examining the emergence of new risks and an increase or shift in existing risks. In traditional industrial installations, there are several protection measures in place, especially modern advanced process control (APC) systems, decompression systems and automatic trigger systems, which prevent risks. However, there continue to be incidents related to the safety of processes. They tend to be more frequent during the start-up phases, as most APC systems are deactivated and factory processes are carried out in manual mode. Alarms are probably deactivated or ignored, as these warnings have been designed to watch for variations in the process in a continuous regime. Sometimes, in order to achieve production, units may function at the limit of the zone of operational functioning, at the threshold for the alarm, before triggering it, which generates further halts and creates more "restarts". A lot of information and data about the process is collected. However, most of them are held in different data silos and are poorly analyzed or integrated in a way that would allow an efficient surveillance of the risks related to process safety. With the rise in new digital technology in the Industry of the Future, these data silos can now be combined and analyzed. This data integration may show, for example, which parts of the factory are vulnerable and subject to more problems or subject to a greater risk. The data may also indicate the set of optimal parameters to avoid any problems, and may help in predicting the next malfunction. Creating virtual environments can also allow operators to acquire practical experience through simulation to identify the right settings for temperature, pressure, flow rate, the position of the valves, etc. These virtual reality methods help operators in making decisions in a variety of scenarios to reduce errors, confusion and risks.

The aim of this book consists of identifying and listing the various attributes and elements that are essential for industrial safety to contribute to raising awareness among the various stakeholders, emphasizing the implication of Safety 4.0 in processes in Industry 4.0.

Chapter 1, "The Industrial Revolution 4.0", begins with a chronological summary of the different industrial revolutions. The definition of the factory of the future is provided. The digital technologies leading to rupture or innovation, related to communication, and the interconnection and management of data from Industry 4.0 are then presented. Finally, the potential impact on the safety of the structure of digital technologies is discussed.

Chapter 2, "The Concept of Safety 4.0", defines the concept of Safety 4.0, examines the history of the evolution of industrial safety and offers a framework for the convergence of the factory of the future and new safety management.

Chapter 3, "Occupational Safety and Health", first identifies the impact of digital technologies on the working conditions of stakeholders. There is then a detailed review of the common and distinct characteristics of health and safety at work, and process safety. This is followed by a list of the different traditional industrial risk analysis methods and techniques. The concepts, paradigms, structural bases and the ways in which they are coupled and the associated complexities are explained in detail. The chapter then comments on the applicability of the methods related to both types of safety (occupational safety and health, and the safety of processes).

Chapter 4, "Process Safety and Cybersecurity", begins by comparing the points of view of cybersecurity and safety processes, respectively. The EBIOS and attack tree risk analysis methods are described. A coordinated approach to reconcile the risk analysis methods for process safety and cybersecurity shows the richness of the synergy and interactions of these methods. Many analogies, like the preliminary hazard analysis and Cyber PHA, the HAZOP and Cyber HAZOP methods, the bow-tie graphs and cyber bow-tie charts, the LOPA and Cyber LOPA methods and the integrated ATBT method, are all highlighted. Finally, it is recommended that it is prudent to use risk matrices and concatenated matrices.

Chapter 5, "Examples: Safety 4.0 and Processes", brings together various examples in a novel way, illustrating the place and influence of Safety 4.0 in the design, implementation, use and reconfiguration and re-designing the processes of the future. The diversity of the 16 examples chosen practically demonstrate the specificity of each approach and the plurality of the digital technologies implemented.

Chapter 6, "Intensification and Inherent Safety: Myth or Reality?", first reviews some essential elements of the intensification of processes. It then describes nine examples of the intensification of processes, systematically highlighting Safety 4.0 aspects. There is an attempt to rationalize and create a general framework for the synthesis and design of intensified equipment, integrating complex indicators and limitations of safety. The myth and/or reality, a priori, of Safety 4.0 are methodically examined using tools and methods dedicated to intrinsic safety. Applying these to six examples, provided in detail, reveals the conflicts within Safety 4.0 with respect to processes, especially when studying the dynamic behavior of intensification processes.

The demands of the challenges related to products and processes in Industry 4.0 simultaneously involve good practices around risk prevention, occupational safety and health, process safety and cybersecurity, as well as social acceptance and environmental responsibility. The contents of this book are meant to promote a reciprocal dialogue between digital technology professionals and actors within the field of industrial safety. The diversity in the many examples provided here should make it possible to look at analogical problems and questions around Safety 4.0 with respect to new processes emerging from the industry of the future. All stakeholders must take charge of the concept of Safety 4.0 and its implementation.

## Acknowledgments

I would like to express my wholehearted gratitude and friendship to Jean-Pierre Corriou, Professor Emeritus at the University of Lorraine, who shared all the scientific and material uncertainties throughout the progression of this book with exceptional availability. His pragmatic and semantic contributions, in response to my endless requests and questions, were very stimulating.

I am very honored that Neil Mitchison acceded to my request to kindly write the preface to this book, for which I thank him very warmly. I was lucky enough to have shared and appreciated his internationally renowned skill and knowledge in the field of safety, within INERIS, in his various professional roles within the European Community in Brussels (Belgium), Edinburgh (Scotland), as well as in the Ispra Joint Research Center (Italy).

I would also like to express my deep gratitude to Roda Bounaceur, Bruno Delfolie, Gérard Verdier and Valérie Warth, engineers working in the Network, Administration,

Computer and Development department of LRGP, whose professionalism in setting up robust hardware and software environments helped me to work remotely.

I would also like to include Laure Thomas-Geoffroy for her resilient assistance in scientific documentation.

I would like to acknowledge the faithful support of my colleagues Laurent Perrin and Olivier Dufaud, Professors of the Process Safety at the University of Lorraine, who welcomed me to the SAFE group.

I could not have completed this work without the constant support of my family.

February 2023

# List of Notations

| | |
|---|---|
| AI | Artificial intelligence |
| AIoT | Artificial Intelligence of Things |
| ALARA | As low as reasonably achievable |
| ALARP | As low as reasonably practicable |
| ANSSI | *Agence nationale de la sécurité des systèmes d'information* – National Agency for the Cybersecurity of Information Systems |
| APC | Advanced process control |
| AT | Attack tree |
| ATA | Attack tree analysis |
| ATBT | Attack tree bow-tie |
| BBN | Bayesian belief network |
| BDMPs | Boolean–logic driven Markov processes |
| BO | *Bulletin officiel* – Official bulletin |
| BPCS | Basic process control system |
| BWM | Best worst method |

| | |
|---|---|
| CCE | Center critical event (top event) |
| CCPS | Center for Chemical Process Safety |
| CHAZOP | Computer HAZards and OPerability analysis |
| CPS | Cyberphysical system |
| CPPS | Cyberphysical production system |
| CSTR | Continuous stirred tank reactor |
| CYPSec | CYber Physical Security |
| DREAL | *Direction Régionale de l'Environnement, de l'Aménagement et du Logement* – French Regional Management of the Environment, Development and Housing |
| DRT | *Direction des relations du travail* – French Regional Management of Labor |
| EAST | Easy attractive social timely (cognition) |
| EBIOS | *Expression des besoins et identification des objectifs de sécurité* – Expression of needs and identification of security objectives |
| EDD | *Étude de dangers* – Study of the hazards |
| EFCE | European Federation of Chemical Engineering |
| EPSC | European Process Safety Centre |
| ERP | Enterprise resource planning |
| ETA | Event tree analysis |
| EV | Source of threats |
| FMEA | Failure mode and effect analysis |
| FMECA | Failure modes, effects and criticality analysis |
| FRAM | Functional resonance analysis method |

| | |
|---|---|
| FTA | Fault tree analysis |
| GTST-MLD | Goal tree-success tree and master logic diagram |
| HAZAN | HAZard ANalysis |
| HAZOP | HAZard and OPerability analysis |
| HF | Human factor |
| HOF | Human and organizational factors |
| HSE | Health, safety and environment |
| HW | HardWare |
| ICCA | International Council of Chemical Association |
| ICEP | Installation Classified for Environmental Protection |
| IChemE | UK Institution of Chemical Engineers |
| IDS | Intrusion detection system |
| IEC | International Electrotechnical Commission |
| IIoT | Industrial Internet of Things |
| ILO-OSH | International Labour Standards On Safety and Health |
| IMPULSE | Integrated Multiscale Process Unit with Locally Structured Equipments |
| INERIS | *Institut national de l'environnement industriel et des risques* – French National Institute for the Industrial Environment and Risks |
| INRS | *Institut National de Recherche et de Sécurité* – French National Institute of Research and Security |
| INSET | INherent SHE Evaluation Tool |
| INSIDE | INherent SHE In DEsign |
| IOHI | Inherent occupational health index |

| | |
|---|---|
| IoS | Internet of Systems |
| IoT | Internet of Things |
| IOW | Integrity operating window |
| IPL | Independent protection layer |
| IS | Information system |
| ISA | International Society of Automation |
| ISC | ICheme Safety Center (UK) |
| ISD | Inherently safer design |
| ISHE | Indicators for safety, health and environment |
| IT | Information technology |
| ITU | International Telecommunication Union |
| L | Likelihood |
| LOPA | Layer of protection analysis |
| LOTL | Living Off The Land (cybersecurity) |
| MASE | *Manuel d'amélioration de la sécurité des entreprises* – Manual to Improve Safety in Companies |
| MES | Manufacturing execution system |
| MoC | Management of change |
| MS | Management system |
| NFR | Non-functional requirements |
| NIST | National Institute of Standards and Technology |
| OECD | Organization for Economic Cooperation and Development |
| OHSAS | Occupational health and safety assessment series |

| | |
|---|---|
| OPR | Open Plate Reactor |
| OSH | Occupational safety and health |
| OSHA | Occupational safety and health administration |
| OT | Operational technology |
| OVI | Operator of vital importance |
| OV | Targeted objective |
| P | Probability |
| PAH | Polycyclic aromatic hydrocarbon |
| PCB | Polychlorinated biphenyls |
| PCRDT | *Programme cadre de recherche et développement technologique* – Framework programmes for research and technological development |
| PEEK | PolyEther Ether Ketone |
| PHA | Preliminary hazard analysis |
| PID | Piping and instrumentation diagram |
| PPE | Personal protection equipment |
| PRA | Preliminary risk analysis |
| PPRT | *Plan particulier des risques technologiques* – Specific plans for technological risks |
| RAMI 4.0 | Reference Architectural Model Industry 4.0 |
| RCPA | *Réacteur continu parfaitement agité* – Perfectly stirred continuous reactor |
| REX | Return of EXperience (feedback) |
| RFID | Radio Frequency IDentification |
| RMM | Risk management measures |

| | |
|---|---|
| RRF | Risk reduction target factor |
| SCADA | Supervisory control and data acquisition |
| SIL | Safety integrity level |
| SIS | Safety instrumented system |
| SL | Security level |
| SME | Small and medium enterprise |
| SMS | Safety management system |
| SR | Source of risk |
| SSM & IFD | Six-step model and information flow diagram |
| STAMP | Systems-theoritic accident modeling and process |
| STPA | Systems-theoritic process analysis |
| SW | SoftWare |
| UD | Unique document |
| V | Vulnerability |
| VOC | Volatile organic compounds |
| WHSV | Weight hourly space velocity |

# 1

# The Industrial Revolution 4.0

Process industries that transform matter into energy, implementing chemical or physical processes, manufacture essential and innovative products that can improve well-being and the quality of everyday life in society.

Despite their indisputable contribution to the advances in standard of living, from the very beginning, all these activities have included intrinsic hazards and potential risks that must be managed. However, the implementation of these risk management measures is a difficult and demanding task. Our vision of this risk must not only be understood and viewed from an industrial or technological point of view, but must also include the choices made by people, citizens and society as a whole.

However, the problem must first be situated within the current industrial context.

## 1.1. A history of industrial revolutions

The various industrial revolutions have always been preceded by scientific, technological and organizational advances and innovations (André 2019). We present a brief overview of these earlier revolutions before introducing Industry 4.0.

Figure 1.1 illustrates the chronology of these different revolutions. It must be pointed out that the exact dates of the transitions related to each can fluctuate a little across literature.

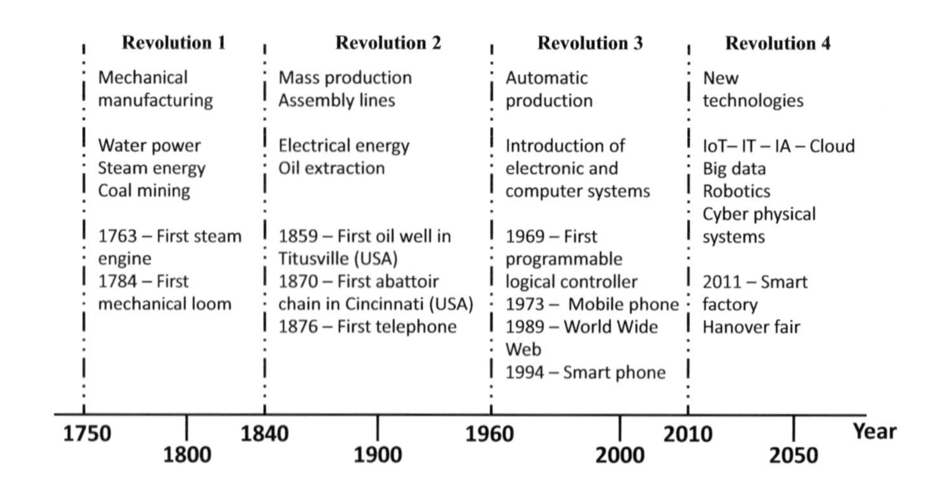


**Figure 1.1.** *The chronology of industrial revolutions*

The first Industrial Revolution, or Industry 1.0, which begins here around 1750, was based on coal mining, metallurgy, the emergence of the weaving industry and the steam engine.

The second Industrial Revolution, or Industry 2.0, which began around 1840, was founded on electricity, oil wells and the birth of the mechanical and chemical industries. The earliest means of communication appeared around this time, with the first operational telegraph line (1833) and the telephone (1876). The railways became a means of public transport. In 1911, F.W. Taylor pioneered the scientific management of organizations. Henry Ford launched the assembly line manufacturing of an automobile.

The third Industrial Revolution, or Industry 3.0, began around 1960, with the emergence of electronics (transistors and integrated circuits), computer science, telecommunications, audiovisual and the nuclear industry. Industrial production was especially impacted by automation and robotics.

The latest and current Industrial Revolution, Industry 4.0, began in 2010. A new cyberphysics system brought together software, sensors and means of communication to manage complexity, anticipate malfunctioning and steer performance in real time. For the first time, resources, information, machines, tools and workers were connected in a network to create an industrial Internet of Things (IoT).

Breque et al. (2021) have already initiated a new transition toward Industry 5.0. According to the authors, Industry 5.0 will recognize industry's capacity to achieve social goals going beyond employment and growth by becoming a resilient source of prosperity. Production will respect the needs of the planet by placing the well-being of stakeholders, including workers, at the heart of production and manufacturing processes.

In France, the new Pacte Law (2019) aims to establish Corporate Environmental and Social Responsibility by creating the status of "Entreprise à mission", a legal framework whereby companies set environmental and social goals that they must achieve. The benefit of this framework is that it allows companies to frame their statutes around a mission made up of a set of freely chosen objectives that work for the greater good. For example, the company Danone, the only one of France's CAC 40 companies to have chosen this status, committed to several goals, including the promotion of best food practices, supporting a better and more sustainable mode of regenerative agriculture, giving each salaried employee the chance to weigh in on company decisions, as well as providing support to the most vulnerable actors in the company's ecosystem. The recent leadership crisis (2021) at the head of this food group will probably allow us to judge the robustness of this sociolegal innovation.

## 1.2. Defining the factory of the future

The technological advances in the fourth Industrial Revolution resulted in a new generation of factories, which have been given various labels: the factories of the future, smart factories, digital factories, cyberfactories, integrated factories, innovative factories, Factory 4.0, and even Industry 4.0. The concept of "Industry 4.0" was born out of a strategic initiative announced by the German government at the Hanover Trade Fair in 2011(Kagermann et al. 2013).

Following a bibliographic analysis, Hermann et al. (2016) considered that factory 4.0 is a collective term denoting the technologies as well as the concept of the organization of the value chain. In the smart factories with a modular structure, in Industry 4.0, cybersystems monitor physical processes, creating a virtual copy of the physical world and taking decentralized decisions.

By using the IoT, cybersystems communicate and cooperate with each other and with humans in real time. Internal and inter-organizational systems have been offered via the Internet of Systems, and these systems are used by members of the value chain.

## 1.3. Technology used in Industry 4.0

To clarify the semantics of the terminology relating to technologies in Industry 4.0, Julien and Martin (2018) have suggested categorizing digital technologies into three categories:

– disruptive technologies;

– technologies for communication and interconnection;

– data management technologies.

### 1.3.1. *Disruptive technology*

A disruptive technology or innovation is an innovation (often technological) related to a product or a service that ultimately replaces an existing technology that had dominated the market thus far. It gives rise to a new category of products or services that had not previously existed.

#### 1.3.1.1. *Additive manufacturing: 3D and 4D*

The NF-E-67-001 standard defines 3D additive manufacturing or 3D printing as the set of manufacturing processes that enable joining materials to create physical objects from 3D model data, layer by layer, as opposed to subtractive manufacturing methodologies.

4D additive manufacturing, or 4D printing, consists of adding a new dimension to 3D printing or 3D manufacturing. The fourth dimension aims to bring in scalable functionalities that evolve over time based on the input of external energy from various sources. This fourth dimension is most often obtained by using smart materials, which most often correspond to active, transformable or programmable hardware. This involves adding information to the material, or endowing it with specific properties so that it can respond to stimuli that are electrical, magnetic, chemical, thermal, vibratory, etc., and can transform itself and cause the product to change (properties, shape, color, conductivity, etc.).

#### 1.3.1.2. *Robotics*

Robotics is made up of the scientific and industrial fields that are related to design, study and creation of robots and their applications. In the industrial domain, robotics results in automatons that carry out precise functions in assembly chains. Robotics also produces devices that can move around in different hazardous environments: polluted, radioactive, aerial, submarine, in outer space, etc. Apart from industries, robotics is also used for scientific research, space exploration, military defense, and maintaining law and order. It is also used in the medical sector for prostheses and assisting healthcare workers. Robotics is also now available to the general public through autonomous devices that carry out specific tasks (vacuum cleaners, lawn mowers) or through entertainment devices (robotic toys).

##### 1.3.1.2.1. Cobot

"Cobot" is a neologism formed from the words "cooperation" and "robotics". The cobot is a small and light robot that works directly with the operators, helping them by carrying out the most thankless and cumbersome tasks. The main distinguishing feature of the cobot is that it interacts with a human, hence the name "collaborative robot". This technology, which is already all the rage within Factory 4.0, allows the operator to gain in productivity and presents absolutely no danger in the workplace. It is especially likely to open up avenues for robotic applications within small and medium enterprises (SMEs).

Blaise et al. (1993) published an interesting guide to better understand what consequences the use of a cobot has on the health and safety of its operators.

In the sixth chapter of their book, Julien and Martin (2018) offer a highly pedagogic introduction to the methodology for using a cobot, illustrated with an example of an end-of-line packaging workstation that has been automated.

##### 1.3.1.2.2. Exoskeletons

Exoskeletons and other physical assistance devices, first developed for the medical sector, are used more and more frequently within companies. They were introduced as systems that made it possible to complement efforts to assist operators. Exoskeletons are defined as external structures worn by the operator, designed to provide physical assistance in carrying out a task. They may be powered (active exoskeletons) or non-powered (passive exoskeletons). These devices make it possible to enhance mobility and, sometimes, even improve physical capabilities. From the point of view of prevention of occupational diseases, it is hoped that these systems will compensate for the efforts put in by operators and thus limit the development of musculo-skeletal problems.

### 1.3.2. *Technologies used for communication and interconnection*

#### 1.3.2.1. *Web and mobile applications*

Devices and solutions in this category include mobile phones, smartphones, tablets, laptops, drones and mobile applications. The use of 5G technology in mobile networks is expected to facilitate and accelerate digitization and the transformation toward Industry 4.0. However, 5G, which is accessible to the general public, has proven to be the continuation of a movement that had already been initiated. The general public is likely to see this new standard as an evolution from 4G, rather than as a revolution. For companies, on the other hand, 5G is a revolution as its more global approach to information system analysis and the solutions that it can offer in the long term bring in potentially significant changes. On the other hand, the fundamental question of long-term cybersecurity of systems is an urgent one, both in terms of development as well as implementation.

#### 1.3.2.2. *The Internet of Things*

The IoT brings together objects and equipment connected to the Internet as well as the technologies used in the network and software related to these devices. The International Telecommunications Union defines the IoT as a "global infrastructure for the information society, enabling advanced services by interconnecting (physical and virtual) things based on existing and evolving interoperable information and communication technologies".

#### 1.3.2.3. *Industrial Internet of Things*

The Industrial Internet of Things (IIoT) is the use of sensors and smart devices to improve manufacturing processes and industrial processes. It is also known as the Industrial Internet or Industry 4.0. The IIoT uses the power of smart machines and real-time analysis to make use of data produced by machines in industrial environments over years.

Although the Internet of Objects and the Industrial Internet of Objects have many technologies in common, especially cloud platforms, sensors, connectivity, machine-to-machine communication and data analysis, they are used for different purposes. IoT applications connect devices in various sectors, especially smart agriculture, healthcare, companies, consumers and public services, as well as city administration. IoT devices include smart accessories, like widely available fitness bands and other applications. On the other hand, IIoT application connect machines and devices in sectors such as oil and gas, public services and the manufacturing industry. System failure or down time in IIoT deployments could lead to high-risk situations or even life-threatening situations. IIoT applications are more concerned with improving efficiency and with health or safety than the user-centered IoT applications.

#### 1.3.2.4. *The Artificial Internet of Things*

The artificial intelligence (AI) of things refers to the combination of the IoT and AI. In other words, the AIoT opens up paths to a new world in which interconnected things can now benefit from AI techniques. Thanks to AI, the IoT has machine-learning capabilities. On the other hand, AI benefits from the information collection and exchange capabilities in the IoT.

#### 1.3.2.5. *Augmented reality*

Augmented reality is a technique that makes it possible to directly visualize objects or physical environments in an enhanced manner, using a digital support (glasses, tablets, smartphones, etc.). The video stream that is displayed retranscribes the physical world by putting in additional information.

#### 1.3.2.6. *Virtual reality*

Virtual reality immerses the user into a virtual or semi-virtual world. The whole of the visualized world is virtual and the user can interact with the components of this world. Additional explanations are often provided in textual or audio form and there may also be a virtual assistant.

#### 1.3.2.7. *Digital twin*

A digital twin is a virtual clone or replica of a physical system or a process. It systematically involves the existence of a pair made up of the digital model and the object that is being copied. The objects involved may be a product, a machine, a production line, a process or a supply chain. Depending on the concerned system and its desired use, this model may be geometric, multiphysical, functional or behavioral. It must evolve over time like its real twin. This twin makes it possible to improve steering, safety and the optimization of production lines and factories, enhance digital continuity at the product level, from design to end-of-life, and enhance monitoring and predictive maintenance. It makes it possible to put in place new economic models in the supply chain. It makes it possible to increase the quality of products by improving process correction. It allows increased traceability of objects and processes, integrating greater information on the components, suppliers and production. It is a disruptive tool when it comes to training needs and demonstrations using systems that are complex and hard to physically duplicate or transport.

#### 1.3.2.8. *Smart product*

A smart product has knowledge, AI and integrated communication capabilities. It knows what it is, what it must do and what it must work on. It may even know its location and what other devices and users are in its vicinity. Endowed with sensors, actuators and ambient intelligence, a smart device may engage with users and other devices through multimodal interfaces, which allows it to communicate and provide advice or suggestions.

### 1.3.3. *Data management technology*

#### 1.3.3.1. *Cloud and cloud computing*

##### 1.3.3.1.1. The cloud

The cloud is defined by a global network of servers, each performing a unique function. This is not a physical entity, but a vast network of inter-connected remote servers around the world, which are meant to function as a unique ecosystem. The servers are applicable through web applications from anywhere on earth (Zwingelstein 2019).

#### 1.3.3.1.2. Cloud computing

Cloud computing is a model that allows anytime, practical and omnipresent access to a shared group of configurable computing resources, such as networks, servers, storage, applications and services, which can be shared with service providers with minimal management. Cloud computing is an infrastructure in which the computing power and storage are managed by remote servers to which users connect through a secure Internet connection. A desktop or laptop, mobile phone, tablet and other connected devices become access points to run applications or to consult data that are stored on these servers. The cloud is also characterized by its flexibility, which allows service providers to automatically adapt the storage capacity and computing power based on the user's needs. The chief advantages of cloud computing consist of the possibility of sharing resources, storing a large amount of data, accessing any service from anywhere and at any time through an Internet connection, and accessing this from any terminal.

The disadvantages that are often highlighted are as follows:

– the platform's safety and security when it comes to preventing the risk of intrusion and data theft through piracy;

– the long-term viability of the host site, given the fluctuation in the service market, with the risk of closure leading to data loss.

#### 1.3.3.2. *Big Data*

"Big Data" refers to all digital data produced by using new technology, whether for personal or professional use. This covers company data (emails, documents, databases, chronological account of data, trades or professions, etc.) as well as data from sensors, Web content (images, videos, sounds, texts), e-commerce transactions, exchanges over social networks, data transmitted by connected objects (electronic stickers, smart counters, smartphones, etc.), geolocalized data, and so on. Given its vast size, Big Data cannot be processed conventionally using computerized tools. The general characteristics of Big Data are traditionally summarized by the 3V principle, which was then broadened to 5V: volume, variety, velocity, value, veracity (Commenge 2020):

– "Volume" refers to the quantity of information acquired, stored, processed, analyzed and shared.

– "Variety" resides in the heterogeneity of formats and data types. These data may be numbers, tables, digital files, characters, texts, images, sounds, videos or all these elementary formats together.

– "Velocity" refers to the dynamic or temporal character of the data and their continuous updating in the form of data flow. This flow can be analyzed either in a deferred manner, if the data are stored and processed in batches, or in real-time, when the processing, distribution and streaming methods are available.

– The "value" of any data is not linked to the value the data takes, nor the financial value of the system used to acquire it, but its potential for valorization, especially in economic terms. For a company or administration, better internal data management may result in reduced costs and greater operational efficiency.

– The "veracity" of data refers to its quality and to the ethical considerations related to the use of these data. Technically, this corresponds to the validity and reliability of the data (aberrant, biased or missing values) and the confidence we have in the system or person generating these data.

#### 1.3.3.3. *Data mining*

Data mining is a field that emerged following the explosion in the quantity of information stored, with significant advances in processing speeds and storage supports. The objective of data mining is to discover useful information, in large quantities of data, which could help in understanding the data or in predicting the behavior of future data. From the beginning, data mining has used statistical tools and AI to achieve its objectives. Data mining has been integrated into knowledge extraction and is a field that is growing by leaps and bounds. It is often defined as being the process of discovering new information by examining large quantities of data stored in data warehouses using shape recognition technology, as well as statistical and mathematical techniques.

This initially unknown information may be in the form of correlations, software design patterns, standardization processes or general trends in the data. Modern science and engineering are based on the idea of analyzing problems to understand the underlying principles and develop appropriate mathematical models. Experimental data are then used to verify the validity of the system or for estimating certain tricky parameters in mathematical modeling.

However, in the majority of cases, the systems have no principles that are acquired or understood, or which are too complex for mathematical modeling. The development of computers made it possible to gather vast quantities of data about these systems. Data mining aims to use these data to create models by assessing the relationships between the input and output variables in these systems. Indeed, everyday thousands upon thousands of data points are produced and recorded by banks, hospitals, scientific institutions, shops, etc.

Data mining represents the entire process using information technology, including the most recent developments, to extract useful information from these data. Various automated tools are used to do this. The procedure begins with a description of the data, summarizing their conventional statistical attributes, visualizes the data using curves, graphs, and diagrams, and finally tries to find potentially significant links between the variables. However, the description of the data itself does not provide an action plan. A prediction model must be built based on the information that was discovered and then this model must be validated on other data (not the original data used to build the model). Today, data mining has considerable economic significance, as it makes it possible to optimize the management of human resources and material resources. For example, it is used in:

– credit companies, to decide whether or not to extend credit to someone based on the applicant's profile, their request and their past experience with loans;

– optimizing the number of vacancies in planes and hotels and for overbooking;

– the layout of shelves in supermarkets to best display products;

– organizing advertising campaigns, promotions and targeting offers;

– medical diagnosis of patient symptoms and pathologies;

– genome analysis;

– classifying objects;

– e-commerce;

– analyzing sales practices and strategies, and their impact on sales;

– Internet search engines with text mining;

– data evolution over time and mining sequences.

#### 1.3.3.4. *Blockchain*

Blockchain is an Internet protocol based on cryptography. It makes it possible to secure digitally transferred data sent from a unique data source to a unique recipient. The proof of the existence of the data source and the consistency and veracity of the transferred data are highly important elements in this protocol.

#### 1.3.3.5. *Cybersecurity systems*

The International Telecommunication Union (ITU) defines cybersecurity systems as "the set of tools, policies, security designs, safety mechanisms, guidelines, risk management methods, actions, training, best practices, guarantees and technologies that can be used to protect the cyber-environment and the assets or organizations and users". Different normalization instances and digital security institutes have proposed norms and guides to help industries in their various processes, which includes the important IEC 62443 standard developed by the International Society of Automation (ISA).

## 1.4. Attempts at structuring technologies

It is instructive to try and structure the technologies listed here to look at their key dimensions and the abilities that are conducive to using them in process industry 4.0. Results from various enquiries carried out among industries to identify the most promising Industry 4.0 technologies are presented in the literature.

For example, Hermann et al. (2016) chose the following four technologies: cyberphysical systems (CPS), IoT, Internet of Services (IoS) and *smart factory*. They used these to derive design principles for scenarios that may arise in Industry 4.0. Table 1.1 presents the six design principles for each of these technologies.

| Technology principle | CPS | IoT | IoS | Smart factory |
|---|---|---|---|---|
| Interoperability | X | X | X | X |
| Virtualization | X | – | – | X |
| Decentralization | X | – | – | X |
| Real-time capacity | – | – | – | X |
| Service orientation | – | – | X | – |
| Modularity | – | – | X | – |

**Table 1.1.** *Design principles based on the selected technologies (Hermann et al. 2016)*

These principles should help companies look for any potential Industry 4.0 pilot projects that could be implemented.

Similarly, given the rapid evolution of Industry 4.0 technologies, Keller (2018) has shown the significance of this temporal evolution. Figure 1.2 thus presents the pertinence of digital technologies and their development. The figure was established based on responses from 10,693 employees working in the German chemical industry. The relevance of 2018 and 2025, respectively, were evaluated on a scale from 0 to 10. For each technology, the number in the circle represents the relative degree of change in the period being considered.

Figure 1.2 shows the significance and notable evolution of three modeling and simulation technologies, cloud and big data. Illustrations from the same presentation are available from the point of view of the following professions: innovation, industrial engineering, purchasing, quality and production, repair and maintenance, and sales and marketing.

Badri et al. (2018) also compared the advantages and drawbacks of six Industry 4.0 technologies in terms of the potential impact on safety. This has been summarized in Table 1.2.

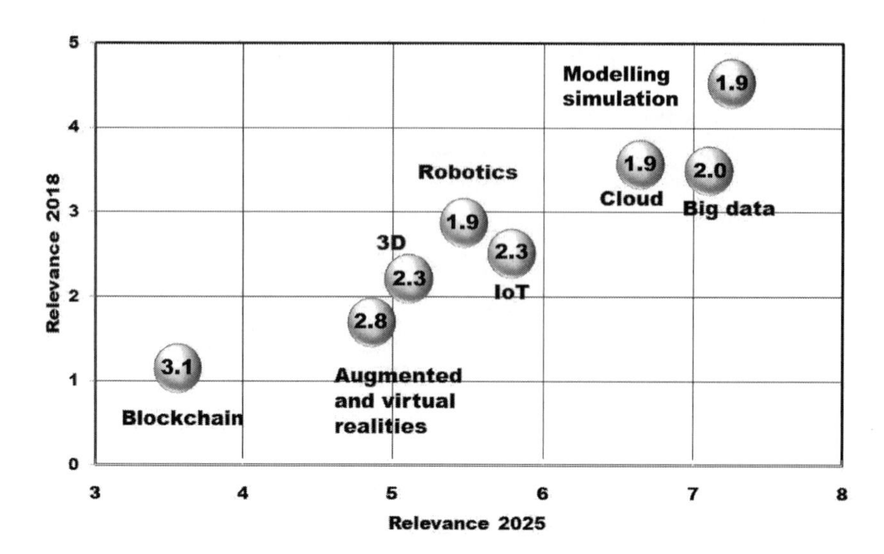


**Figure 1.2.** *Pertinence of digital applications and their development (Keller 2018)*

| Technology | Advantages | Drawbacks |
|---|---|---|
| Big Data | - Unlimited gathering of data<br>- Reduced uncertainty<br>- Enhanced capacity for behavioral analysis<br>- Anticipation of errors | - Data reliability<br>- Data selection criteria<br>- Personal data confidentiality |
| IoT<br>CPS | - Improved interaction between equipment/machinery and anomaly detectors | - Network reliability<br>- Cybersecurity |
| Cobotics | - Improved flexibility and accessibility | - Unpredictability of worker reliability, proximity and interaction with devices |
| IA | - Learning and quick recognition of hazards<br>- Real-time decision and action | - Uncertain reliability<br>- Potential drift (calibration)<br>- Absence of standards |
| Simulation | - Improved evaluation and comparison of work scenarios and methods<br>- Risk prevention at source | - Uncertain reliability and robustness of the models |

**Table 1.2.** *Advantages and drawbacks of the six chosen technologies in terms of the impact on safety (Badri et al. 2018)*

Forcina and Falcone (2021) recently published a systematic analysis of literature to show how Industry 4.0 technologies can benefit from enhanced safety management. The authors, who examined 68 articles published between 2010 and 2020, presented several examples and reported that the two most widely used digital technologies for safety management were the IIoT and the cloud.

The use of smart devices, advanced monitoring systems and the development of digital information are all likely to involve new competencies with respect to existing best practices.

Finally, Thienen et al. (2016) proposed a multi-tier architecture, presented in Figure 1.3, to help industries engage in the Industry 4.0 process.

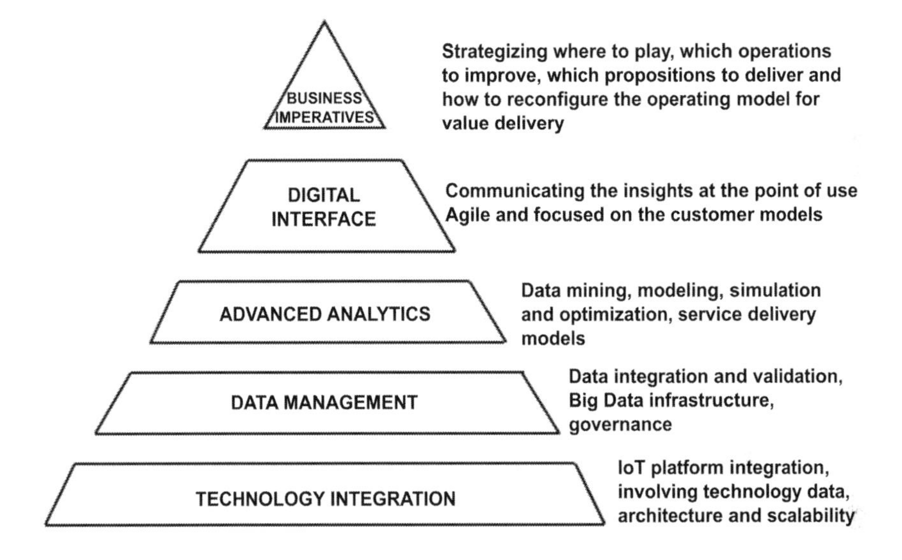


**Figure 1.3.** *The architecture of the structure of solutions, dimensions and capacities conducive to process industry 4.0 (Thienen et al. 2016)*

The goal of this architecture is to allow the company to build a digital network containing sequences that bring together abilities in the different domains required for digital transformation.

The tiers in this structure successively describe, starting from the base, technological integration, data management and advanced analysis. In the physical domain, these smart assets are manifested in the layer of digital interfaces used to steer digital capabilities and, finally, the company's strategic requirement.

## 1.5. Conclusion

In summary, defining Industry 4.0, the panorama of technologies that make up Industry 4.0, their importance and temporal relevance form the framework for industrial safety, which must adapt to new demands.

# 2

# The Concept of Safety 4.0

## 2.1. Context and definition

Pasman and Fabiano (2021) report that the neologism "Safety 4.0" was introduced in the 16th EFCE Symposium International on Loss Prevention held in 2019 in Delft, The Netherlands. In terms of industrial safety, the new challenge for process industries, namely transforming the factory of the future, is represented by the absolute need for a transition between Industry 3.0 and Industry 4.0. New approaches are required to reorganize process safety and the manner in which technological risks must be managed.

The concept of "Safety 4.0" is not limited only to digital security as a new paradigm in the era of Big Data and Industry 4.0, but also includes a complete conceptual and methodological transition as well as a new paradigm shift. On the one hand, it is essential to understand the potential effects of technological changes on employment and the individual health and well-being of the workers as well as on performance and the impact it has on the organizational level, identifying strategies to increase confidence and the acceptance of disruptive digital technologies. On the other hand, the health and safety implications of Industry 4.0 technologies in terms of organization of work, the regulatory and legislative framework, safety management systems for processes and occupational hazard management systems must be analyzed in detail.

Based on this, Laciok et al. (2021) defined "Safety 4.0" as proactive change in the science of process safety and work, which focuses on the resilience of systems and dynamic risk management based on four pillars, namely interoperability (IoT), information transparency (digital twin), technical assistance and decentralized decision making.

## 2.2. The history of the evolution of safety

As with the chronology of the different industrial revolutions, we present below the history of the evolution of industrial safety, revisiting the same transition dates. Figure 2.1 illustrates the succession of different safety sequences.

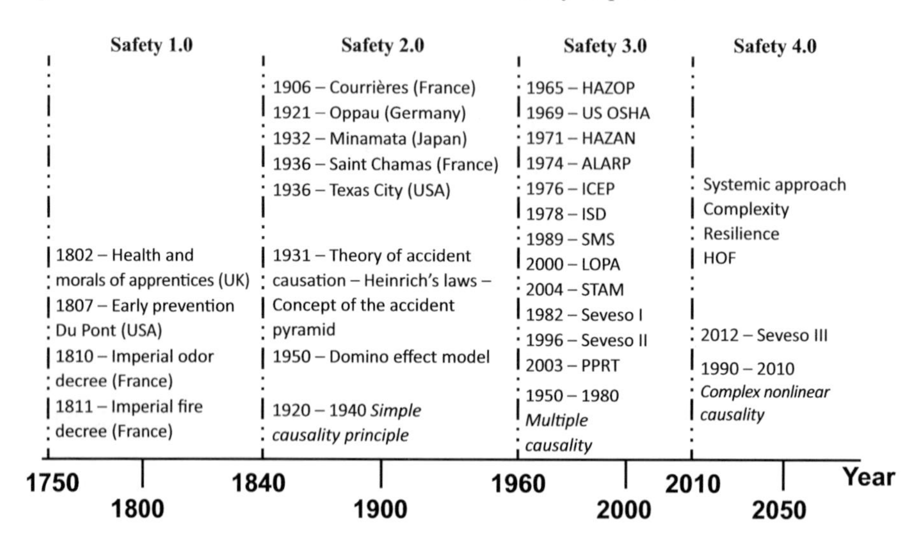


**Figure 2.1.** *Historical overview of the evolution of safety*

Prescriptions for safety began in the "Safety 1.0" period, with the first law on "Health and morals of apprentices" published in the United Kingdom in 1802. Imperial decrees published in France in 1810 and 1811 formed the earliest bases for regulations on anthropic risks.

The imperial decree dated October 15, 1810, touched upon manufacturing and workshops that released insalubrious or uncomfortable odors. This decree identified three categories of establishment based on the nature of activity carried out:

– eleven industries were judged to be compatible with urban installations and were placed in the first category, such as gilding or soap-making industries;

– twenty-three others, placed in the second category, required only simple administrative monitoring of the harmlessness of their methods, such as candlemakers or white lead manufacturers as well as foundries, including lead smelters;

– considered to be dangerous or insalubrious, only the 31 activities listed in the third category had to carry out their activities away from habitation, such as chemical soda ash plants or manufacturers of varnish, minium, ammonium salt, sulfuric acid or Prussian blue.

The imperial decree dated September 18, 1811, led to the creation of the Paris Fire Brigade to fight against conflagrations. In a way, these texts were the precursors to regulations around future classified installations.

The "Safety 2.0" sequence was marked by significant major accidents, such as the Courrière mine disaster (1906) – the largest mining disaster in Europe – explosions in fertilizer factories in Oppau (1921) and Texas City (1947), explosions in gunpowder plants, including Saint-Chamas (1936), and the Minamata health disaster due to mercury poisoning (1932). In terms of a paradigm, the theory of a single causation of accidents, the Heinrich laws, the pyramid concept of accidents and the domino effect were successively published.

The "Safety 3.0" stage saw an abundance of new risk analysis tools, such as the failure tree analysis (FTA) developed in 1962 by Bell Laboratories (USA). Then there were the HAZOP, HAZAN, layer of protection analysis (LOPA) and STAMP methods, the ALARP concepts, safety management systems (SMS) and inherently safer design (ISD), and the French ICEP environmental protection system for classified installations. A series of major technological disasters still punctuated these 50 years. Among others, there were disasters in Feyzin (1966), Flixborough (1974), Seveso (1976), Los Alfaques (1978), Three Mile Island (1979), Mexico (1984), Bhopal (1984), Basel (1986), Pasadena (1989), La Mède (1992), Enschede (2000), Toulouse (2001) and Buncefield (2005). Given the severe consequences that were seen, regulations grew exponentially stricter. The European directives, Seveso I and II, were promulgated in 1982 and 1996, respectively, and then turned into national legislation by various member states. In 2003, following the explosion in a factory storing ammonium nitrate, in Toulouse (2001), France brought in an important law on *Plans particuliers des risques technologiques* (PPRT) (Specific Plans for Technological Risks). Finally, between 1950 and 1980, the theory of single causation for accidents was widened to multiple causation for accidents.

The start of the new "Safety 4.0" era is linked here to the concept of the smart factory, introduced by the German government at the Hanover trade fair. New digital technologies used in Industry 4.0 are increasingly disrupting traditional learning in industrial safety. New major technological accidents still occurred between 2010 and 2020, including disasters following accidents in Fukushima (2011), Rouen (2013 and 2019), Tianjin (2015), Yancheng (2019) and Beirut (2020). The Seveso III directive was published in 2012, becoming effective in 2014. The systemic or inter-organizational approach results in nonlinear complex causal modalities being taken into account. It is now inevitable that human and organizational factors as well as the concept of resilience must be taken into consideration.

## 2.3. Safety framework

This section aims to offer a framework for considering safety in general in the new environment of Industry 4.0. A review of open literature shows the large number of articles published on the theme of Industry 4.0.

However, most articles chiefly only mention the roles of new digital technologies as drivers of the Industrial Revolution 4.0. There are still only a few texts that examine the introduction of safety in the context of Industry 4.0, from the occupational safety and health point of view as well as the point of view of process safety.

Liu et al. (2020) have proposed a framework that integrates the concepts of Industry 4.0 and Safety 4.0, which they consider to be a new safety management paradigm.

At its core, Figure 2.2, which illustrates the proposed canvas to work on, aims to work toward the convergence of the factory of the future and safety management 4.0. The upper part summarizes the five new digital technologies and four new design principles, which act as support and guide. The diagram in the lower half of the figure brings together the principle, system and means of managing safety focusing on advanced and smart digital aspects.

Putting this framework into practice is not easy a priori. Indeed, what should be understood by the term "safety"? France, like most countries, is yet to arrive at a consensus on its definition, its meaning or its acceptance between different stakeholders in the fields of occupational safety and health (OSH), process safety and the cybersecurity of industrial systems.

Nonetheless, there is movement on integrating different points of view and this is encouraged and sometimes even demanded. For example, the renewed collaboration between the teams for the inspection of classified installations (DREAL) and work inspection teams is governed by the DRT instruction of April 14, 2006 (DRT 2006).

Consequently, and despite the fact that the processing methods used by different entities for safety may often be very specific, Chapter 3 will look at the OSH aspect without neglecting the influence of process safety. Similarly, Chapter 4 will describe similarities in the joint approaches of process safety and cybersecurity.

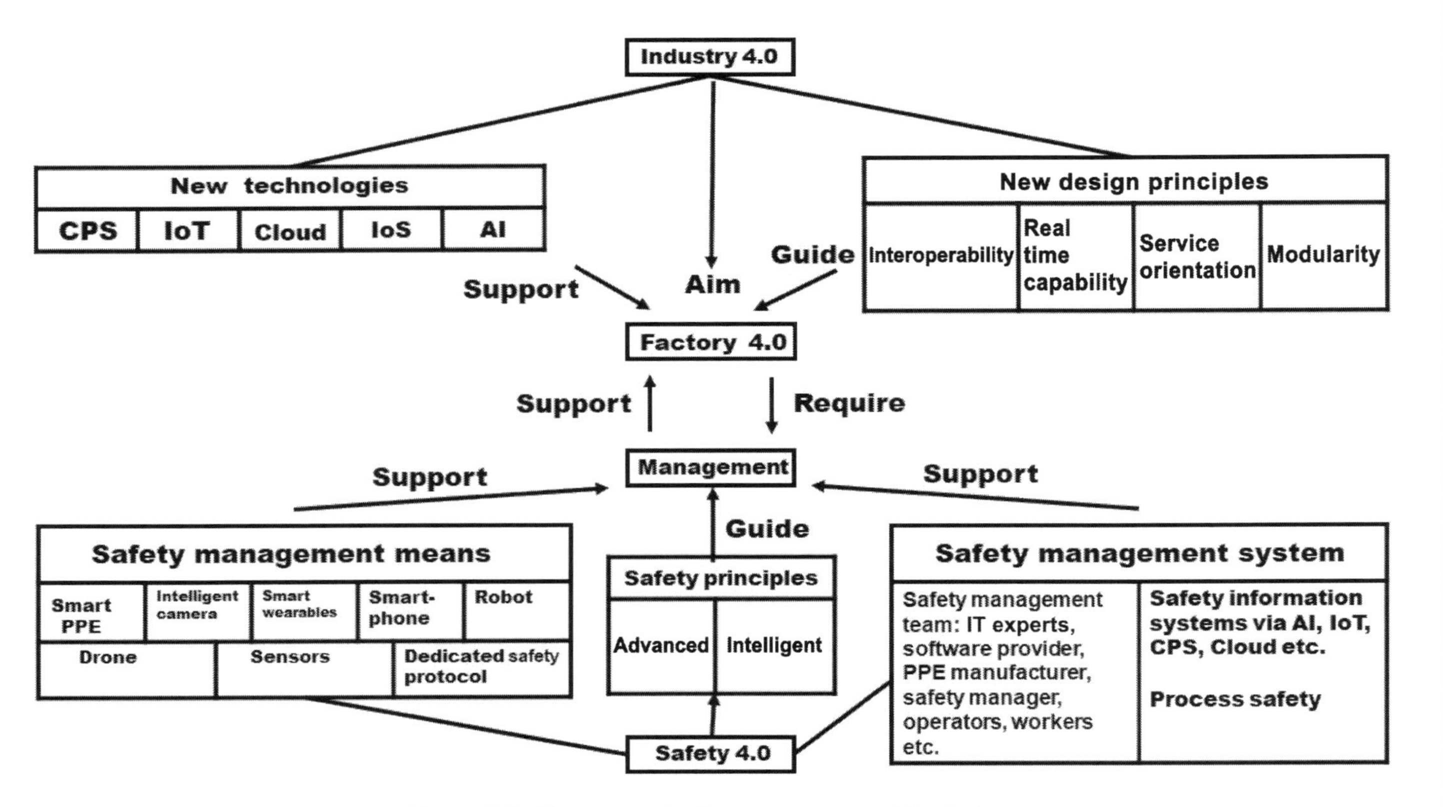


**Figure 2.2.** *Framework for the convergence of the factory of the future and safety management 4.0 (Liu et al. 2020)*

# 3

# Occupational Safety and Health

This chapter first identifies how Industry 4.0 digital technologies will impact work conditions for the stakeholders. It then goes on to examine how the emergence of Industry 4.0 makes it necessary to revisit and simultaneously adapt risk analysis methods for safety and health at work as well as process safety.

## 3.1. Impact of Industry 4.0 work conditions

Leso et al. (2018) have listed the chief opportunities and advantages from the application of Industry 4.0 in workplaces. Work conditions and the organization of work will change. Employees will continue to play a key function in retaining knowledge about their work, including decentralized decision-making activities and the evaluation of the quality of production processes. This may also mean that these works will be involved in more activities that are more creative, interesting and which add value. Thus, they will be able to qualitatively enrich their work, step away from repetitive tasks and achieve greater autonomy and personal development. It is important to note that Industry 4.0 involves greater organizational complexity, which will require flexible work conditions. These conditions can offer employees a better work–life balance, as well as offer better compatibility between personal development and continued professional development.

In addition, the flow of information along the supply chain could make industrial management more transparent and better organized, thus reducing the hierarchical pressure on the workforce. Industry 4.0 could make work safer and healthier through early and continuous risk analysis and management, based on intelligent safety technologies and virtual engineering. Monitoring technologies, such as portable technologies (e.g. helmets and bracelets fitted with sensors), can help employees stay safe within dangerous work environments, where they may be exposed to extreme heat, toxic gases, open flames, or hazardous chemical products. This technology must allow a continuous monitoring of the employee's health and well-being (e.g. sudden anomalies such as cardiac arrest, a fall, or progressive changes in stress levels) and also be able to monitor the state of the equipment, the machines and installations.

Further, when interpreting what has been monitored and by then selecting the appropriate action to perform, self-adaptive and self-learning machines, endowed with advanced analysis capacities, may be able to anticipate dangerous situations during worksite operations. They can then use predictive algorithms and health-management algorithms to manage unexpected conditions, thereby preventing accidents and injuries for the workforce as well as for passersby. The technological capabilities of Industry 4.0, associated with cognitive analysis, could improve employee capabilities by facilitating comprehension and promoting employee safety and well-being. Industry 4.0 will benefit from the availability of functional industrial robots to carry out a growing number of tasks, including painting, soldering and assembly, which may benefit from the robot's strength, endurance and precision. In this way, productivity and quality could increase progressively, while at the same time musculoskeletal problems, and traumatic or fatal injuries, as well as the costs of products and services, could be reduced or avoided altogether. Furthermore, operator health and safety could be ensured by using professional robots for dangerous operations in disaster zones. This type of robot generally interacts unidirectionally with the human operator who controls it and sends this operator information about its environment and its functions. For the record, cobots, or collaborative robots, have been developed to directly interact with human workers equipped with robotic devices, thereby improving performance.

Globally, this kind of advanced automation makes work environments more ergonomic and comfortable. Physical assistance devices can also act as mobile helpers and provide physical support, for example, prostheses, robotized exoskeletons and personal porter robots. Exoskeletons have been developed to help workers in carrying out tasks such as lifting and manipulating weights, which could increase their productivity. These devices could, therefore, help in offering jobs that are more flexible and inclusive at a social level to a workforce that is increasingly diverse in terms of age, sex and cultural context. It could also help workers who were injured during their rehabilitation period. In this way, Industry 4.0 could allow people to continue working and remain productive for longer via flexible and customized careers.

In summary, Leso et al. (2018) highlight that there is a better interface between work and private life, work is less monotonous, there is better human–machine interaction for dangerous tasks, working conditions are safer and cleaner, and there is tracking of the health and well-being of employees and smart personal protection equipment (PPE). However, Leso et al. (2018) do not shy away from listing the chief problems and preoccupations in the workplace, namely, the increased psychological risks, the impact on private life, the decrease in human contact, the need for a qualified workforce and the uncertain question of unemployment. In the same vein, looking at the problem along the axes of changing job markets, social values and the type of work (manual or cognitive), Schulte et al. (2020) and Medhammer et al. (2021) reported, in an analysis of potential scenarios and risks in Industry 4.0, that there was a large prevalence of psycho-social risks to employees at work. The resilience of the group of older, anxious workers is also problematic when it comes to adaptation and reconversion.

## 3.2. Definitions

Occupational safety and health (OSH) is a multidisciplinary approach that has the objective of removing or reducing the risk of accidents that could occur when performing some professional activity. Occupational hazards could be the result of carrying out some activity that has not been mastered, cramped postures or the use of toxic chemical products. Broad categories of occupational hazards include risks related to physical activity, the risk of falls or slips, the risk of infection, psychosocial risks, risks related to atypical work timings and risks related to the commute to and from the work place. These risks may result in bodily and/or psychological harm and lead to occupational diseases with immediate or delayed effects.

Process safety consists of managing the entirety of systems and business units by applying inherently safer design principles, flexible engineering and rigorous good operating practices. It deals with prevention, mitigation, attenuation and protection during incidents or accidents that could result in a dangerous loss of control of equipment or energy. Such a loss of control could have serious consequences, such as a fire, an explosion and/or toxic effects leading, finally, to extensive damage to property, a significant impact on the environment and leading to production halting or slowing down, with financial losses and negative effects to brand image and reputation.

## 3.3. OSH versus process safety

Kerin (2019) considers that three key features distinguish OSH from process safety:

– the causal mechanisms: while process safety and OSH are both concerned with a potential, dangerous loss of control over energy, process safety generally concerns the management of higher levels of energy;

– the severity of potential consequences: while there may be fewer incidents with process safety than OSH incidents, their consequences are likelier to be serious;

– the emphasis on engineering and design: process safety is concerned with the safety of the industrial system, while OSH is oriented toward the safety of personnel interacting with the system.

The inability to identify these differences and to develop appropriate management practices has been an important factor in many industrial safety incidents and accidents.

As shown in Table 3.1, the European Process Safety Centre (EPSC 2021) has also summarized some of the differences between the domains of OSH and process safety.

The ISC (Icheme Safety Center) of the Institution of Chemical Engineers[1] has listed a few distinct and shared features of the roles of OSH professional and process safety professionals, respectively (Table 3.2).

## 3.4. OSH assessment of occupational hazards

### 3.4.1. *Regulations, norms and unique document*

The working of an OSH management system may be based on an explicit frame of reference, although a company is not obliged to choose an existing frame of reference. Several standards and references have been published successively, such as the MASE (1990), ILO-OSH (2001), OHSAS 18001 (2007) and the current ISO 45001 (2018) standard.

ISO 45001 (ISO 2018), "Occupational health and safety management systems: Requirements with guidance for use", specifies the requirements for an occupational safety and health management system and provides guidelines for how this should be used so as to allow organizations to procure worksites that are safe and healthy by preventing trauma and pathologies related to the work, and by proactively improving their OSH performance. This international standard is applicable to any organization that wishes to establish, run and keep up to date an OSH management system with the aim of improving health and safety at work, removing dangers and minimizing OSH risks (including system failures), leveraging OSH opportunities and remedying any non-conforming features in the OSH management system that are related to the organization's activities. The standard aims to help an organization achieve the desired

1 Available at: https://www.icheme.org/media/14929/0007_18-competency_brochure-update.pdf.

results with its OSH management system. In accordance with the organization's OSH policy, the desired results for an OSH management system include:

– the constant improvement of OSH performances;

– satisfying legal requirements and other requirements;

– achieving OSH objectives.

| Element | Life Saving Rules - OSH | Process Safety |
|---|---|---|
| Objective | Reduce number of injuries/fatalities | Avoid loss of containment of chemicals with potentially serious consequences for people, environment and assets |
| HSE domain | Behaviors in occupational safety | Behaviors in operators involving hazardous chemicals |
| Target population | All | Operation teams on hazardous sites |
| Nature and applicability | Simple rules that are easy to understand and apply in all circumstances | More complex principles that cannot always be fully applied (e.g. in case of design issues) |
| Implementation | Non-negotiable set of requirements (Life Saving Rules / Golden Rules) | The aim is to identify situations that are not in line with the Process Safety Fundamentals and to start a discussion on how to proceed, while avoiding uncontrolled initiatives "to get the job done" |

**Table 3.1.** *A few differences between OSH and process safety (EPSC 2021)*

ISO 45001 applies to all organizations, regardless of size, status and activity. It is applicable to all OSH hazards that are under the control of that organization, taking into account factors such as the context within which the organization has developed, as well as the needs and expectations of its workers and the other shareholders. This standard does not define specific criteria for OSH performance, nor does it give specifications for the design of an OSH management system. It allows a company to integrate other aspects of health and safety, such as well-being and quality of work at life through the OSH management system. The standard does not deal with product safety, material damages or environmental risks, apart from risks to workers and other concerned parties. The ISO 45001 can be used in its entirety or in part to systematically improve OSH management. The French National Institute of Research and Safety (Institut National de Recherche et de Sécurité - INRS) has published many articles and recommendations to help companies adopt OSH management processes (Aubertin 2007; Aubertin et al. 2007; Drais 2018; Hoguin 2018)[2].

2 Available at: https://www.inrs.fr/demarche/evaluation-risques-professionnels.

| Element | Process Safety | Scope | OSH |
| --- | --- | --- | --- |
| Focus | Approach focused on high-consequence, low-frequency issues resulting in loss of control with potentially catastrophic consequences | Public and environmental impact of the operations | Main focus on workers, the impact of the processes on people<br>Emphasis on management systems |
| Risk management | Hazard identification based on detailed, systematic analysis.<br>Risk assessment focus on operational risks associated with the process and equipment.<br>Quantitative risk assessment.<br>Risks to community, workers, and the facility | Similarity in fundamentals of hazard identification and risk assessment<br>Concept of the hierarchy of control.<br>Warning signs of potential loss of control.<br>Awareness of consequences of loss of control.<br>Semi-quantitative risk assessment<br>Risk to the environment | Hazard identification based on a range of information, including consultation with stakeholders.<br>Workplace risks associated with the work undertaken by people or that impacts people<br>Qualitative risk assessment processes |
| Emergency preparedness | Predictive analysis, e.g. consequence modelling<br>Focus on containing the process | Preparedness of the system's response.<br>Environmental impact of emergencies and emergency response; recovery after emergency | Focus on personal safety<br>Hazard-specific quantitative risk assessment<br>Risk to workers |

| Element | Process Safety | Scope | OSH |
|---|---|---|---|
| Engineering and design | Design and hazard analysis to inform and support inherently safer process plant | Plant/operator interface | Structures, materials and plant/equipment with an emphasis on plant life cycle and worker safety |
| Asset integrity - inspection and maintenance | Integrity of critical controls<br>Equipment reliability | Condition monitoring | Inspection and maintenance schedules |
| Management of change (MoC) | Engineering and technical change, temporary design or operational changes<br>Consistent, up-to-date documentation | Resolution of potential issues from changes to plant, equipment, process or people<br>Managing people through change via communication and consultation | Changes having an impact on the organization of work, the environment or standards impacting work; may be organizational, legislative or other sources |
| Systems and procedures | – | Systemic and systematic management approach | – |
| Safety systems analysis | Evaluation of process safety MS effectiveness and reliability of barriers Process safety performance metrics | Systems review | Evaluation of OSH MS effectiveness and risk controls OSH performance indicators |
| System manuals and drawings | Accuracy of the technical information and drawings | Documentation review | Currency of documentation relating to to worker safety |
| Root cause analysis | – | Systematic analysis processes | – |
| Contractor and supplier selection and management | – | Contractor competence | Contractor personnel safety |

| Element | Process safety | Scope | SST |
|---|---|---|---|
| Reporting and investigation | Reporting of process deviations | Legal requirements for reporting<br>Analysis to identify trends<br>Return of Experience (REX) | Incident and injury reporting |
| Legislation, regulation, codes and standards | Focus on specific duties assigned within the legislative requirements for high-hazard activities | Environmental legislation | OSH specific legislation |
| Audit, assurance, management review and intervention | Audits of asset integrity against engineering standards | Management system audits | Hazard and compliance audits on plant, equipment, chemicals, asbestos, training, housekeeping procedures and behaviors |
| | Continuous review focuses on systemic root causes | – | Improvement processes focus on both immediate and latent causes |
| Human factors | Impact of the person on the process and integrity of the system | Interaction of the person, task and organization | Impact of the process on the person |

**Table 3.2.** *Common and distinct features of the roles of OSH professionals and process safety professionals (source: ISC IChemE)*

The “Unique Document” (UD) for the assessment of occupational hazards is a mandatory document for all companies. This must list all occupational hazards that could be encountered by workers and the related actions for prevention and protection. The UD is one of the pillars of the OSH management systems, as the OSH policy, and the organization that follows from this policy, is based on this document. It is established in seven steps:

1) Initial analysis: establish a preliminary inventory of risks and the preventive actions that have already been undertaken.

2) Prevention policy: the process must be based on a genuine desire, on the part of the head of the organization, to commit to a continuous process of improvement and advancement for the company. The objectives that are set must be appropriate for the size of the company and the nature of its activities, and this policy must be clearly explained, defined and communicated to the employees.

3) Organization: a clear and precise designation and explanation of the roles played by different actors in the company; training, informing and consulting employers and their representatives.

4) Planning actions: identifying and assessing hazards using a participative method. Establishing objectives for the prevention of each risk. Drawing up a plan of actions defining concrete measures to be implemented to attain these objectives.

5) Implementation and functioning: ensuring there is monitoring of the implementation of the planned prevention measures. Ensuring regulations are respected, including temporary workers or sub-contractors.

6) Measuring performances, analyses and corrective actions: verifying the efficacy of the implementation and reacting as soon as a new hazard is identified. Maintaining an OSH dashboard with indicators (frequency, severity of work accidents) and analyzing work accidents with or without injuries.

7) Improving the management system: carrying out regular audits to improve the functioning of the system, if required, or to enhance the company’s performance. There must be regular updates during management reviews where all managers are present. The three objectives consist of modifying the policy, if required; of drawing up new objectives and targets, and tracking the various results obtained (audits, action plans) in order for the system to keep evolving.

Implementing these seven steps ensures the continuous improvement of the management system, which will ensure that it continues to become more efficient (Anteol et al. 2004; Labbe 2006).

How can we reconcile the prevention of occupational hazards and the industry of the future? The employer must involve all concerned employees (operators, management, workers collectives) as far upstream as possible, in order to identify viable solutions that are best suited to their actual work, to anticipate the training that they will require, and to promote the acceptance of these technological and organizational changes. As with any work equipment, the use of new technology must be preceded by a complete risk assessment. Finally, the steps for prevention must be capable of being adapted in a relevant and effective manner to potentially rapid and frequent changes. The prevention of occupational hazards in Industry 4.0 must focus on three major transformations:

– advanced production technology that relies on greater digital integration and the use of innovative materials and processes (additive manufacturing, collaborative robotics, etc.);

– digitization, leading to the interconnection of all elements in the manufacturing process (operators, products and machines) and to the opening of these to the internet;

– flexibility, with the aim of obtaining production systems that can be rapidly reconfigured based on client requirements.

Special attention must be paid, in the unique document, to potential psycho-social risks produced by these transformations, which have already been highlighted in section 3.1. Ravallec et al. (2013) and Langevin and Guyot (2020) have proposed documents that could help in assessing psychosocial risk factors.

### 3.4.2. *Inventory of risk analysis techniques and methods*

The ISO 31010 standard (ISO 2019) describes the principles of risk management as well as the processes and organizational frameworks related to this field. It specifies a process to recognize, understand and, if necessary, modify the risk based on the criteria established within the framework of this process. The methods and techniques used to assess the risk could be applied in the context of this structured approach, involving the establishment of a context, the assessment and handling of the risk with monitoring, review, communication, concerted efforts, consignment and continuous recording.

The risk assessment methods and techniques that are described in the standard can be used:

– when it is necessary to better understand the hazards present or a particular hazard;

– in the context of a risk management process that gives rise to actions to handle the hazard;

– in the context of a decision where multiple options, each involving risks, must be compared or optimized.

In particular, these methods and techniques can be used to:

– supply structured information that supports decisions and actions in case of uncertainty;

– bring in clarity about the consequences of hypotheses about achieving the objectives;

– compare several options, systems, technologies, approaches, etc., when each object is subject to multi-faceted uncertainty;

– facilitate the definition of realistic operational and strategic objectives;

– contribute to determining risk criteria for an organization (risk limits, the appetite for risk or risk-taking capacity, for example);

– taking into account the risk when defining or revising priorities;

– recognize and understanding the risk, including a risk that could have extreme consequences;

– understand the essential uncertainties around the organization's objectives and justify what must be done about them;

– recognize and make use of opportunities more successfully;

– clearly articulate the factors that contribute to the risk and the reasons why they are important;

– identify the effective and efficient handling of risks;

– determine the modifying effects of the proposed methods for handling the risk, including all modifications relating to the nature or severity of the risk;

– communicate about the risk and its implications;

– learn from successes and failures to improve how the risk is managed;

– demonstrate that regulatory and other requirements have been respected.

The ISO 31010 standard specifies that the manner in which the risk is to be assessed depends on the complexity and novelty of the situation, as well as the corresponding level of knowledge and understanding. In the simplest case, if a situation presents no new or unusual elements, if the risk is well understood, and if no stakeholder has intervened or if the consequences are not significant, actions can be decided based on established rules and procedures, and on previous assessments of that risk. For unprecedented situations, which are complex or difficult, if there is high uncertainty and limited experience, there is insufficient information to form the basis of the assessment, and conventional analysis techniques may prove to be useless or irrelevant. This may apply equally to situations where stakeholders have diametrically opposite points of view. In these cases, several methods and techniques may be used to obtain a partial understanding of the risk by forming judgments within the context of organizational and societal values and by taking into account the different points of view of the stakeholders.

The new digital technology in Industry 4.0 does indeed give rise to new risk-assessment situations that are complex and difficult. In Chapter 1, section 1.4, there was a list of all the non-negligible assets of Industry 4.0: real-time communication, big data, human-machine cooperation, monitoring and control, autonomous equipment and inter-connectivity. By restricting themselves to questions related to the human operator in a sociotechnical context, but excluding big data and real-time communication, Adriaensen et al. (2019) transposed the four principles into four key challenges: interconnectivity, autonomy of machines, automation of human behavior conjoined with their professional role and a change in supervision control. This latter factor is derived from tele-detection of monitoring and controls. Based on this, the authors established an inventory of different traditional risk-analysis methods and techniques by defining the concepts, the paradigms, the bases for structuring and the nature of the coupling and complexity of the concerned systems (Adriaensen et al. 2019; ISO 2019). This is summarized in Table 3.3.

### 3.4.3. *Applicability of risk analysis methods to OSH*

It is clearly necessary to carry out a critical analysis of the various traditional risk-analysis techniques and methods. The brief explanation of the concepts behind these methods is enough to identify the content of each. It is thus possible to note that the concept behind certain methods is focused on:

– identifying the hazards;

– assessing the risk or analyzing the probability of an accident;

– the assumed sources of the hazard;

– analyzing deviations;

– the interactions between systems.

Studying the basis of structuring of these methods shows that there is considerable variation. For example, the structuring of these techniques can be classified based on:

– the propagation of events and failures;

– the presence of barriers and defenses that protect against this failure propagation;

– the existence of assumed sources of hazards;

– the analysis of deviations;

– the presence of interactions between systems.

| Method | Concept | Paradigm | Basis for structure | Coupling/Tractability |
| --- | --- | --- | --- | --- |
| Energy analysis | Identifies energies that can harm human beings | Energy barrier thinking | Volumes that jointly cover the entire object | Loose coupling - tractable |
| Hazard and Operability Studies (HAZOP) | Identifies deviations from intended design of equipment based on the use of predetermined guide words. It is generally carried out by a multi-disciplinary team during a series of meetings | Linear causality | Deviation of operational parameters | Tight coupling - tractable |
| Failure Mode and Effect Analysis (FMEA) | Identifies failures of components and their effects on the system | Linear causality and decompositional analysis | Reliability from components or modules | Tight coupling - tractable |
| Fault Tree Analysis | Causal factors are deductively identified, organized in a logical manner and represented pictorially in a tree diagram that depicts causal factors and their logical relationships to the top event | Energy barrier thinking, linear causality and decompositional analysis | Fault propagation resulting from initial event | Loose coupling - tractable |
| Event (Effect) Tree Analysis | Analyzes alternative consequences of a specified hazardous event | Energy barrier thinking, single cause philosophy and decompositional analysis | Fault propagation back to the initial event | Loose coupling - tractable |
| Action Error Method | Identifies departures from specified job procedures that can lead to hazards | Taylorism | Phases of work of operator | Loose coupling - tractable |

| Methode | Concept | Paradigm | Basis for structure | Coupling/Tractability |
|---|---|---|---|---|
| Job Safety Analysis | Identifies hazards in job procedures | Rationalist, prescriptive and top-down belief in procedures and Taylorism | Elements in an individual job task | Loose coupling - tractable |
| Deviation Analysis | Identifies deviations from the planned and normal production processes | Rationalism, prescriptive and top-down belief in procedures and Taylorism | Activities (e.g. activity flow or job procedure) | Loose coupling - tractable |
| Safety Function Analysis | A structured description of a system's safety functions, including an evaluation of their adequacies and weaknesses | Energy barrier thinking | Defenses or safety functions of the system | Loose coupling - tractable |
| Change Analysis | Establishes the causes of problems through comparisons with problem-free situations | Failure without acknowledging context sensitivity or emergent behavior | Discrepancy between as-is and as-should-be situations | Loose coupling - tractable |
| Root Cause Analysis (RCA) | Attempts to identify the roots or original causes instead of dealing only with immediately obvious symptoms | Single cause philosophy, linear causality and decompositional analysis | Initiating failure causes and effects | Loose coupling - tractable |
| Human Reliability Assessment (HRA) | Identification and prediction of human errors in relation to strictly predefined tasks | Human reliability assessment and Taylorism | Human error | Loose coupling - tractable |
| Deterministic Probabilistic Risk Assessment | Deterministic probabilistic risk assessment (PRA) produces a semi-quantitative measurement of risks based on frequency and severity scales | (Semi-)quantitative causality credo | Ordinal or cumulative frequency and/or severity of harmful events | Loose coupling - tractable |
| Databases | Analysis of consequences of chemical risks like fire, explosions, the release of toxic gases or the determination of toxic effects or combinations of chemicals | Database | Chemical and physical reactions | Loose coupling - tractable |
| Cognitive Task Analysis | An analysis method that addresses the underlying mental processes that give rise to errors | Task analysis as the key to understanding system mismatches | Tasks | Loose coupling - tractable |

| Method | Concept | Paradigm | Bases for structure | Coupling/Tractability |
|---|---|---|---|---|
| Bayesian Networks | A method that uses a graphical model to represent a set of variables and their probabilistic relationships. The network is comprised of nodes that represent a random variable and arrows that link parent nodes to child nodes. | Epidemiological causation model | Events (and their related degree of belief) | Tight coupling - tractable |
| Layer Protection Analysis (LOPA) | LOPA is a semi-quantitative method for estimating the risks associated with undesirable events or scenarios and the presence of sufficient measures to control them. A cause-consequence pair is selected and the preventive layers of protection are identified | Epidemiological causation model and energy barrier thinking | Multiple defenses | Tight coupling - tractable |
| Bow-tie Analysis | A simple, diagrammatic way of describing and analyzing the pathways of a risk from causes to consequences. The focus of the bow-tie is on the barriers between the causes and the risk, and the risks and consequences | Epidemiological causation model and energy barrier thinking | Multiple causes and defenses | Tight coupling - tractable |
| STAMP MIT | Creation of a model of the functional control structure for the system in question by identifying the system-level hazards, safety constraints and functional requirements | Feedback control system | Most basic element in the model is a constraint, whereas basic structuring is the feedback control system | Tight coupling - intractable |
| FRAM | Systemic analysis of complex process dependencies, based on the idea of resonance arising from the inherent variability of everyday performance | Functional resonance | Dependencies among functions or tasks | Tight coupling - intractable |
| Event Analysis of Systemic Teamwork (EAST) | A means of modeling distributed cognition in systems via three network models (i.e., task, social and information) and their combination | Propositional network | Task, social and information network connections | Tight coupling - intractable |

**Table 3.3.** *Inventory of different risk analysis methods and techniques (Adriaensen et al. 2019)*

In terms of paradigms, most of the methods listed here are based on the principle of linear causality, except for the bow-tie analysis, layer of protection analysis (LOPA) and Bayesian networks, which apply the principle of epidemiological causality.

The column titled "Coupling/tractability" merits particular attention, especially with respect to the meaning of "tractability". Systems may be described in terms of their coupling being weak/loose or strong/tight. A system's coupling refers to whether the sub-systems and/or components are connected and interdependent from a functional point of view. Consequently, systems with tight coupling are characterized by the following elements (Hollnagel and Speziali 2008):

– buffers and redundancies are deliberately included in the design;

– processing delays are impossible;

– sequences are invariable;

– substitutions in supplies, equipment, personnel are limited and are anticipated in the design;

– there is very little room for maneuvering in terms of supplies, equipment and personnel;

– there is only a single method to achieve the goal;

– tightly coupled systems are difficult to control as an event that occurs in one part of the system propagates rapidly to the other parts.

The majority of the traditional risk analysis methods and techniques are identified to have loose coupling, except for methods that use HAZOP, FMEA, LOPA, bow-tie analysis and Bayesian networks, which have tight coupling.

The term "tractability" has been the subject of much debate in literature. At present, terms like "information" and "description" that are used to qualify complexity have been replaced by the term "tractability".

In practice, this involves attempting to describe how a method may manage or easily control a system's behavior. A system or process is said to be tractable if its operating principles are known, if the corresponding descriptions are simple and not intricate, and especially if the system's behavior remains stable during the description.

Conversely, a system or process is said to be intractable if its operating principles are only partially known or even unknown, if the descriptions are elaborate with many details, and if the system can change before the description is complete. The inventory of the methods and techniques listed here shows that almost all of them are tractable, except for the last three on the list, STAMP, FRAM and EAST, which are intractable.

In light of these observations, it is not easy to conclude which approach to recommend. It must first be recalled that even the ISO 31010 standard leaves it up to each user to decide which risk analysis method, or combination of methods, to use.

Lyu et al. (2019) have tried to compare certain risk assessment methods for physical cybersystems based on process safety requirements, information technology safety requirements, and an integrated approach to safety (Table 3.4). The solution (or solutions) must make it possible to use system-based approaches to analyze and handle increasingly tightly coupled sociotechnical systems, which are becoming harder to manage (i.e. intractable), related to Industry 4.0 processes and technologies. The majority of the traditional methods and techniques studied do not have the ability to carry out a comprehensive analysis of sociotechnical systems.

| | **FTA** | **FMEA** | **HAZOP** | **GTST-MLD** | **STPA** | **ATA** | **CYPSec** | **STPA-sec** | **BDMPs** | **NFRs** | **BBN** | **STAP-Safesec** | **SSM & IFD** |
|---|---|---|---|---|---|---|---|---|---|---|---|---|---|
| Quantitative risk assessment | x | x | | x | | x | | | x | | x | | x |
| Qualitative risk assessment | x | x | x | x | x | x | | x | x | x | x | x | x |
| Mode-based approach | x | x | | x | | x | | | x | | x | | x |
| System-based approach | | | | | x | | | x | | x | | x | |
| Deductive analysis | x | x | | x | x | x | | x | | | | x | x |
| Hierarchical analysis | | | | x | x | | | x | | | | x | x |
| Dynamic analysis | | | | | | | x | x | x | | x | x | |
| Ability to address threats and hazards | | | | | | | | | | | x | | |
| Subjective analysis | x | x | x | x | x | x | x | x | x | x | x | x | x |

**Table 3.4.** *Summary of part of some existing methods and approaches (source: Lyu et al. (2019))*

Nonetheless, Pasman et al. (2017) have proposed a structured approach that makes it possible to simultaneously implement several methods or techniques, whose application to the industrial context, in a prospective or retrospective point of view, must still be validated. In the same vein, Pasman and Knegtering (2013) have considered that taking into account the dynamic evolution of systems in certain traditional methods could allow advances, despite the limitations of their causation models. For example, the graphical model of dynamic Bayesian networks could be integrated with the bow-tie diagram in a dynamic systems analysis (Duval et al. 2007; Khakzad et al. 2013).

Pasman et al. (2013), Hettinger et al. (2015) and Dallat et al. (2019), respectively, have recommended the prevalence of system-based methods integrated with holistic and socio-technical risk analysis systems. Of the methods and techniques listed here, STAMP, FRAM and EAST are intractable methods that respond to the demands of Industry 4.0 in terms of interconnectivity and complexity. Although non-intuitive, these newer methods may still be used as a primary approach without a detailed knowledge of their methodologies. However, they require many resources and are time intensive with respect to the conventional methods. They must make it possible to identify priorities that require more in-depth analyses and also to identify strengths and weaknesses in the models that could lead to more efficient solutions. Adriaensen et al. (2019) have described the contents of three methods and highlighted the challenges related to applying each of these. By prioritizing sociotechnological issues that affect human beings, these authors have successively examined the impacts of interconnectivity, the autonomy and automation of the human-operator's activity, and the evolution of supervisory control.

# 4

# Process Safety and Cybersecurity

The cybersecurity aspect discussed in this chapter is based on the results of the research carried out on the Cybersecurity of Industrial Systems, led by Jean-Marie Flaus, Professor at the Université Grenoble Alpes (Abdo and Flaus 2015; Abdo 2017; Abdo et al. 2017, 2018; Masse et al. 2018, 2019; Flaus 2019).

Risk control and process safety management is related to the International Electrotechnical Commission (IEC) 61508 and IEC 61511 standards. The cybersecurity of industrial systems is governed by the IEC 62443 standard, which provides a framework to design a system to manage the cybersecurity of an industrial installation and cyber-physical systems. The principles behind both these approaches are very similar, and their application must be constructed in a conjoined and simultaneous manner. This integrated complementarity is necessary to ensure the proper, overall functioning of a system where conventional analysis of process safety, with respect to technical, human and organizational hazards, is complemented by risk analyses related to cyber-attacks on units with programmed devices. Table 4.1 compares the points of view of safety process and cybersecurity, based on the characteristics studies (Flaus 2019).

## 4.1. Reviewing risk analysis methods in process safety: example of the bow-tie method

The process described in the standards IEC 61508 and IEC 61511 is implemented during a stage where there is a risk assessment and the allocation and specification of instrumented systems, whose implementation and use are then monitored. The risk assessment involves using one or more risk analysis methods. Table 3.3 summarized an important list of methods, specifying the concept, paradigm, structuring base and coupling-tractability for each. Debray (2006) and Laurent (2011) have provided a detailed handbook of the most widely used methods, including the preliminary risk analysis, HAZOP, the fault-tree method, layer of protection analysis (LOPA) and the bow-tie method. Although the latter is recommended by administrations, the manager of the industrial unit retains the right to choose their own risk analysis method. However, they must then justify its use during the hazard study.

Since the bow-tie method is widely used, it would be useful to review the principle on which it is based. The bow-tie method combines a fault tree analysis and an event tree analysis, arranged upstream and downstream, respectively, of a "Central Critical Event" (see Hourtoulou and Salvi (2003); Salvi and Bernuchon (2003); Salvi and Debray (2006); Delvosalle et al. (2006)). Since the failure tree analysis is based on an inverse deductive method and the event tree analysis is based on a direct inductive method, the bow-tie method brings together both an inductive and deductive approach. Figure 4.1 depicts one representation of accident scenarios based on the generic bow-tie diagram.

| Characteristics | Process Safety | Cybersecurity |
|---|---|---|
| Type of risk | The risk is due to a random, rather rare failure of a component<br>Simultaneous failures unlikely, except for common causes<br>Remote attack almost impossible | The risk is due to a deliberate attack, systematically exploiting one or more vulnerabilities with intent<br>The hacker seeks to exploit one or more breaches simultaneously<br>Easy remote attack |
| Analysis method | Identification of a critical event, then analysis of the causes and consequences. Methods: PHA, AMDEC, fault-tree analysis | Identification of safety needs and critical event, impact and threat analysis. Methods: EBIOS, OCTAVE, CORAS, attack tree |
| Examples of measures | Safety barriers, redundancy, reduction of common modes, training, work organization | Data flow filtering (firewall, diode), secure architecture, IDS, user training and authentication |
| Risk level measurement | Probability and severity of consequences | Likelihood and severity of consequences |
| Specification of the level of tolerated risk | Quantitative SIL level | Qualitative SL level |
| Examples of similar measures | Safety barriers, redundancy, reduction of common modes, training, work organization | Data flow filtering (firewall, diode), secure architecture, IDS, user training, and authentication |
| Examples of potentially conflicting measures | Update to be minimized to reduce risks | Frequent updates to limit vulnerabilities |

| Characteristics | Process Safety | Cybersecurity |
|---|---|---|
| Examples of potentially conflicting measures | Encryption can be a problem for real-time responses. Safety is improved by fall-back positions, degraded operating modes and emergency stops | Encryption improves security. Fall-back positions, degraded operating modes and emergency stops can be exploited by a hacker to compromise the installation |
| Return of Experience | Many REX sources. Future failures are similar to past failures | The REX must be organized in a global way to secure the exploited breaches |
| Distribution of activities among the participants | High independence constraint: designer, auditor, etc. | There is nothing to prevent design and validation being carried out by the same person |
| Regulatory issues | Law 2003-699 of 30 July 2003. Decree 2010 ICEP. Adoption of the Seveso directives | NIST Directive transposed into French law on 26 February 2018 for OVIs |
| Legal aspects | Responsibility can be established based on the physical chain (suppliers, maintenance) | It is often difficult to find the source of attacks which may come from a country where legal action is difficult |
| Certification (Standards) | The Standard IEC 61508 | Competiting standards ISO 27000, IEC 62443, etc. |

**Table 4.1.** *Comparison of the characteristics of process safety and cybersecurity (after Flaus (2019))*

Table 4.2 explains acronyms used in Figure 4.1, supplemented by definitions and specific examples.

The central point of the bow-tie, the central critical event, denotes either a loss of confinement in the case of a fluid, or a loss of physical integrity for a solid (such as decomposition). The left of the bow-tie is a failure-tree that makes it possible to identify causes that generate the central critical event. The right part is an event tree, which determines the consequences of the central critical event. The safety barriers are represented by thick vertical bars, illustrating their role in preventing the accident scenarios from unfolding. It is therefore possible to visualize each potential pathway for an accident scenario from its initial causes to the final consequences it has on the targets identified.

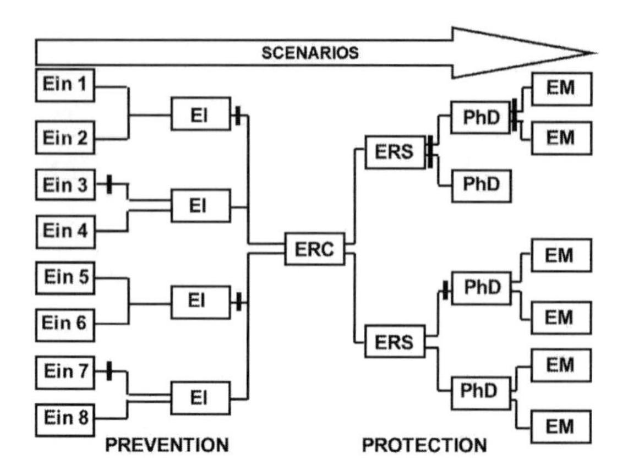


**Figure 4.1.** *Generic representation of accident scenarios based on the bow-tie model (Laurent 2011)*

## 4.2. Risk-evaluation matrix in process safety

Appendix V in the order of May 10, 2000, proposed a model for an initial grid presenting potential accidents in terms of the probability–severity of consequences on humans, as required by the EDD (study of the hazards) (Voynet 2000). Since the circular of May 10, 2010, summarizing the methodological rules applicable to hazard studies, to the assessment of steps to reduce risks at source and to Plans for the Prevention of Technological Risks in classified installations, through the application of the law of July 30, 2003, concerning assessment criteria for steps to control the risk of accidents that could occur in "Seveso establishments" (BO MEDD no. 201/12 dated July 10, 2010), the French administration has proposed a more elaborate grid to assess the probability and severity of consequences. This is a $5 \times 5$ grid, subdivided into 25 cells (Michel 2010). Table 4.3 depicts the *complete* assessment matrix proposed, marking out three areas of accidental risk:

– a *high* risk zone (risk deemed unacceptable), symbolized by the word "NO";

– an *intermediate* risk zone (as low as reasonably practicable [ALARP]), symbolized by the acronym "RMM" (risk management measures), in which the steps for continuous improvement are particularly pertinent, with the aim of achieving as low a risk level as possible, within acceptable economic conditions; this must take into account the state of knowledge, practices and the vulnerability of the environment of the installation;

– a *smallest* risk area (risk deemed acceptable), which does not include "NO" or the "RMM" box.

| Symbol | Meaning | Definition | Examples |
|---|---|---|---|
| EIn | Adverse event | Deviation or failure outside the defined normal operating conditions | Overfilling or fire in the vicinity of hazardous equipment can initiate adverse events |
| EC | Current event | Admitted event occurring on a recurring basis in the life of an installation | Testing, maintenance or equipment fatigue are current events |
| EI | Initiating event | Direct cause of loss of containment or physical integrity | Corrosion, erosion, mechanical aggression, pressure rise, are initiating events |
| CCE | Central critical event or top event | Loss of containment on hazardous equipment or loss of physical integrity of a hazardous substance | Rupture, breach, collapse or decomposition of a hazardous substance in the case of a loss of physical integrity |
| ERS | Secondary critical event | A direct consequence of the CCE, the ERS characterizes the source term of the accident | Formation of a pool or cloud when a two-phase substance is released |
| PhD | Hazardous event | Physical phenomenon that can cause major damages | Fire, explosion, dispersion of toxic cloud |
| EM | Significant effects | Damage caused to targets (people, environment, equipment) by the effects of a hazardous event | Lethal or irreversible effects on the population Accident synergies |

**Table 4.2.** *Inventory of events on the bow-tie model (Laurent 2011)*

The subjectivity of what is "acceptable" must be noted, especially when human life is involved. Indeed, the distinction between the different risk areas is not an easy one and essentially depends on the definitions of the different levels of probability and severity that make up the matrix criticality (Iddir 2008).

The gradation into "NO" and "RMM" corresponds to a risk that increases from level 1 to level 4 for the "NO" boxes and from level 1 to level 2 for the "RMM" boxes. This gradation corresponds to the priority that can be accorded to reducing risks by first striving to reduce the largest risks at the highest levels.

It is useful to spend a little time looking at potential situations that can be envisaged from the probability–severity combination of hazardous events and accidents identified in the EDD (study of the hazards).

| **Probability (a)** | **E** | **D** | **C** | **B** | **A** |
|---|---|---|---|---|---|
| Probability (a) Severity (a) | $\leq 10^{-5}$ | $10^{-5}$ to $10^{-4}$ | $10^{-4}$ to $10^{-3}$ | $10^{-3}$ to $10^{-2}$ | $\geq 10^{-2}$ |
| Disastrous | partial NO (b) RMM level 2 (c) | NO level 1 | NO level 2 | NO level 3 | NO level 4 |
| Catastrophic | RMM level 1 | RMM level 2 (c) | NO level 1 | NO level 2 | NO level 3 |
| Important | RMM level 1 | RMM level 1 | RMM level 2 (c) | NO level 1 | NO level 2 |
| Serious | | | RMM level 1 | RMM level 2 | NO level 1 |
| Moderate | | | | | RMM level 1 |
| Note | Designation | | | | |
| (a) | Probability (per unit and per year) and severity are assessed in accordance with regulations | | | | |
| (b) | New establishments have technical risk control measures in place so that the probability level is maintained in this class when, in each scenario leading to this, the probability of failure in the highest RMM level that could prevent this scenario is taken to 1 | | | | |
| (c) | Existing establishments, if applying for an AS permit extension or modification, check specific criteria from the circular dated 10 May 2010 | | | | |

**Table 4.3.** *Example of a complete probability–severity matrix (Michel 2010)*

– *Situation 1: One or more accidents have a probability–severity pairing that corresponds to a "NO" box.*

The following conclusions can be drawn from this:

– For a new authorization, the risk is assumed to be too large to be able to authorize the installation in its current state; it would be better to ask the owner to modify their project so that the risk is brought down to a lower level. The objective here is to move out of the "NO" boxes.

– For an existing installation, duly authorized, it would be useful to ask the owner for proposals to implement complementary risk reduction measures at source to be carried out within a deadline defined by an order from the local authorities. The aim is to move out of the "NO" boxes, along with conservative measures taken on a temporary basis. If, despite these additional measures, at least one accident remains within a "NO" box, then based on the Prefect's assessment, the risk may cause the closing of the installation by decree of the State Council, unless the additional measurements, taken within a specific regulatory framework, such as a technological risk prevention plan (TRRP), make it possible to move the set of accidents outside the "NO" boxes within a defined deadline.

*– Situation 2: One or more accidents have a probability–severity combination that corresponds to an "RMM" box and no accident falls within a "NO" box.*

It is enough to verify whether the owner has analyzed all foreseeable risk management measures and implemented those whose cost is not disproportionate with respect to the expected benefits, whether in terms of global safety or in terms of the safety of each of the targets present in the environment.

Further, if the total cumulative number of accidents that fall within the level 2 RMM boxes is greater than 5 for the whole establishment, then it must be considered as a global risk equivalent to an accident falling with a level 1 NO box (situation 1). However, if the number of accidents is greater than 5, the probability level for each accident stays in the same probability class when, for each scenario leading to this accident, the probability of failure in the highest risk management level that could prevent this scenario is taken to 1. This criterion is equivalent to considering that the confidence level is brought to zero for this risk management measure (barrier). In practice, this criterion is only possible for accidents in the probability class E.

*– Situation 3: No accident falls within an NO or RMM box.*

The residual risk, considering the risk management measures, is moderate and does not involve any obligation to further reduce the risk of an accident for classified installations.

## 4.3. Risk analysis methods for industrial information systems: example of the EBIOS and attack tree method

In a manner similar to process safety, the information security risk may be defined by the emergence of an uncertain event or a scenario of combined events that could have damaging consequences on a system. The general procedure proposed in the ISO 27005 standard is applicable to different cybersecurity risk analysis methods. The most-widely used methods are listed in Table 4.4.

The most widely used method in France, recommended by the National Agency for Industrial System Safety, is the EBIOS method (Expression des Besoins d'Identification des Objectifs de Sécurité) (EBIOS 2010; ANSSI 2014). The most recent version of this method is the *EBIOS Risk Manager* (EBIOS 2018).

Figure 4.2 presents the five-step process of the EBIOS method, as per the ISO 27005 standard (EBIOS 2018). The essence of this iterative process is carried out within five workshops:

| Name | Designer | Reference |
| --- | --- | --- |
| EBIOS | ANSSI | (EBIOS 2010, 2018) |
| MEHARI | CLUSIF | (Roule 2010) |
| OCTAVE | Carnegie Mellon | (Alberts et al. 1999) |
| CORAS | UiO / SINTEF | (Lund et al. 2010) |
| MAGERIT | The Spanish Ministry for Public Administration | ENISA 2016 European Agency for Cybersecurity |
| SP 800-30 | NIST | (Nurliyani et al. 2019) |
| CRAMM | Central Computer and Telecommunications Agency of the United Kingdom | (Yazar 2021) |
| TARA | MITRE | (Wynn 2014) |
| CVSS | Common Vulnerability Scoring System | (Sheehan et al. 2019) |
| FAIR | Factor Analysis of Information Risk | (Wangen et al. 2018) |
| CMMI | Capability Maturity Model Integration | (Chaudary and Chopra 2017) |
| IoTMM | IoT Maturity Model | (Almajali et al. 2019) |
| OWASP | Open Web Applications Security Project | (Kellezi et al. 2019) |

**Table 4.4.** *Principal methods for cybersecurity risk analysis*

– Workshop 1: framework and core of safety. This defines the scope of the study and the professional and technical boundaries, identifies hazardous events and their degrees of severity and determines the safety base using the references and guides applicable to the study.

– Workshop 2: sources of risk. This identifies and characterizes the risk sources (RS) and the planned objectives. The pairs of RS and objectives are then assessed so that the most pertinent are retained.

– Workshop 3: strategic scenarios. Like in a preliminary risk analysis, this determines the attack pathways that a risk source could take to achieve their objective. The aim is to construct strategic scenarios.

– Workshop 4: operational scenarios. Based on the strategic scenarios constructed in step 3, this step concretizes the development of operational development and assesses the likelihood of each. It is recommended that each scenario be represented in the form of a graph or attack tree.

– Workshop 5: handling risks. After a synthesis of the risks studied, using a risk-matrix assessment, a strategy for risk management measures must be defined, involving a plan for safety measures, estimating residual risk and monitoring how risks evolve.

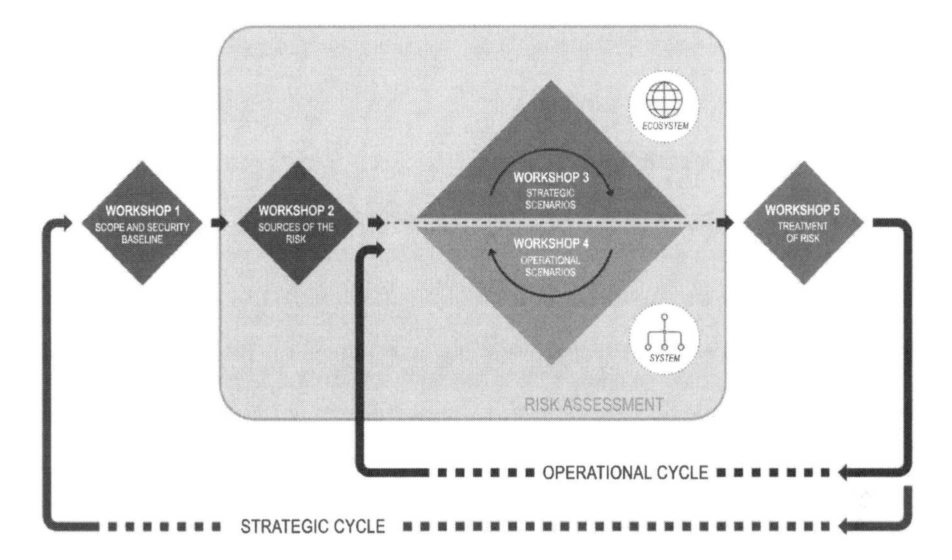


**Figure 4.2.** *Organigram of the steps in the EBIOS method (EBIOS 2018)*

The detailed, step-by-step process to follow for this is explained in some previous works (EBIOS 2018, 2019).

The attack tree, which is in fact identical to a failure tree, is a logic diagram that describes the sequence of events, in series or parallel, that result in the final, undesirable event called the top event. The attack tree is formed of successive levels of events, such that each event is managed from the events on the lower level through AND or OR logic gates. As with the bow-tie diagram, it is possible to materialize defense barriers on the connected arc between two elements.

The "who – how – why and when" methodology is often used to construct a complete attack tree in five steps:

– Step 1: identify all potential attackers, such as competitors, discontented employees, unqualified persons using pirating tools (*script kiddies*) and so on.

– Step 2: determine the plausible goals of each actor in the threat in the first step and create a source for each objective.

– Step 3: reflect on the means that can be used to achieve the main goal. This is the step where we must be creative and envisage all possibilities. Through this process, sub-objectives or steps on the path to the main goal must also be identified.

– Step 4: repeat step 3 for sub-objectives.

– Step 5: review the entire tree and consider the probability of each. The elements that must be considered are the difficulty of the sub-objectives, the time needed, and the skills required for each potential attack.

Consider the simple operational scenario of the source-target system schematized in Figure 4.3(a) using the "Swiss Cheese" model. By moving from the source toward the target, the path taken by the hazard encounters three consecutive layers, whose vulnerabilities are V1, V2 and V3, respectively. They can be corrupted by the threat sources Ev1, Ev2 and Ev3. The path drawn in a thick line illustrates the attack vector. The corresponding attack tree is represented in Figure 4.3(b). The top event consists of the target being attacked. The intermediate and lower level events are interconnected by AND gates.

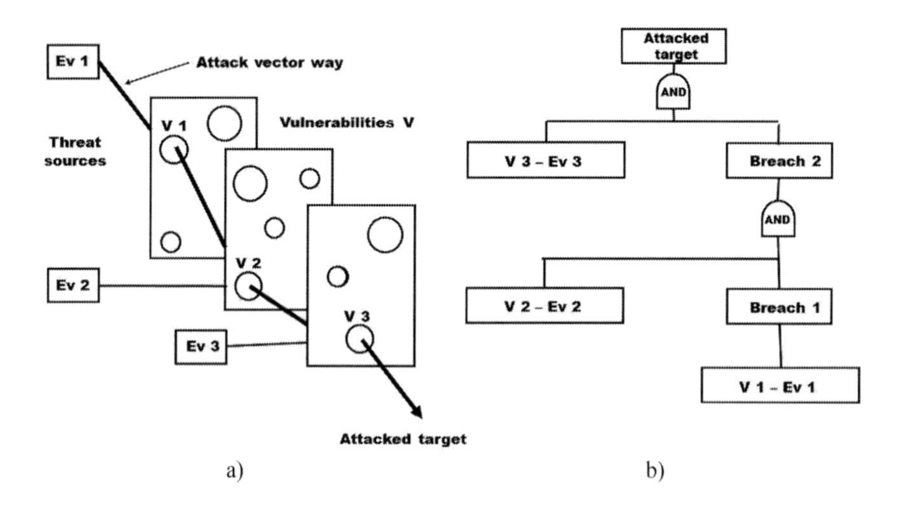


**Figure 4.3.** *(a) The Swiss Cheese model of an operational scenario; (b) example of the corresponding attack three*

The application of graph theory makes it possible to extend the concept of attack tree to the concept of the attack graph representing multiple scenarios, simultaneously generating multiple attack trees. For example, Guan et al. (2011) established the digraph for a SCADA system for a continuous process involving the distillation column for chemical products.

## 4.4. Cybersecurity risk-assessment matrix

ANSSI proposed a simple method to list industrial systems in three categories based on their safety requirements. This classification may be applied to a site, in its entirety, to a more specific section, or to an industrial system spread out over many sites. It is thus recommended that the precise perimeter of the industrial system be defined. Each class must necessarily integrate the measures for the previous class. A succinct description of the three cybersecurity classes for industrial systems is given as follows (ANSSI 2014):

– Class 1: these are industrial systems where the risk or impact of an attack is low. The set of recommended measures for this class must be capable of being completely autonomously applied. This level is the default level for any installation run according to best practices.

– Class 2: these are industrial systems where the risk or impact of an attack is significant. There is no administrative control for this class of industrial systems, but the person in charge of the industry must be able to show proof of the implementation of appropriate measures for control or in the case of an accident.

– Class 3: these are industrial systems where the risk or impact of an attack is critical. In this class, there are stronger obligations and the conformance of these systems is verified by the administration and/or an accredited organization.

These cybersecurity classes have been determined based on impact and vulnerability. The scale of impact (severity) conventionally has five levels based on the human, environmental and economic consequences of the shutdown of the service being delivered (Table 4.5 (Flaus 2019)).

Assessing the likelihood is more difficult. The ANSSI has proposed a detailed, empirical procedure to calculate the likehood (ANSSI 2014). The level of likelihood is first a function of the level of exposure, which results from its levels of functionality and connectivity. The likelihood also depends on the category of legitimate and illegitimate participants in the system, as well as the level of the attackers. Appendix A3 of the proposed procedure (ANSSI 2014) provides an interesting example of the application of this method to a Seveso installation with a continuous production process of toxic chemical products. The qualitative scale of likelihood resulting from the application of the ANSSI method is given in Table 4.6.

Flaus (2019) proposed the simpler Table 4.7 which summarizes the likelihood.

The final matrix of classes is illustrated by Table 4.8.

Analogous to the process safety risk assessment matrix (Table 4.3), it is possible to identify different regions with different risk levels:

– low zone (green), where the risk is judged to be low and acceptable, with no additional prevention, mitigation or protection measures;

– moderate zone (yellow), where the hazard is monitored by indicators and actions that are implemented as far as possible;

– large zone (orange), where the hazard is monitored by indicators and actions that are quickly implemented;

– critical zone (red), where the risk is critical, requiring the shutdown of the installation and the immediate implementation of safety measures.

| Level | | Human Consequences | Environmental consequences | Consequences on the service |
|---|---|---|---|---|
| 1 | Insignificant | Reported accident without stopping or medical treatment | Limited and temporary violation of a rejection standard without legal reporting requirements to the authorities | Heavy impacts on 1,000 people |
| 2 | Minor | Reported accident with sick leave or medical treatment | Violating a discharge standard requiring reporting to authorities but without environmental consequences | Heavy impacts on 10,000 people. Disruption of the local economy |
| 3 | Moderate | Permanent disability | Moderate pollution limited to the site | Heavy impacts on 100,000 people. Temporary loss of major infrastructure |
| 4 | Major | One death. Permanent disability | Significant pollution or pollution external to the site Evacuation of people | Heavy impacts on more than 1,000,000 people Permanent loss of a major infrastructure |
| 5 | Catastrophic | Several deaths | Major pollution with lasting environmental consequences external to the site | Heavy impacts on 10,000,000 people Permanent loss of critical infrastructure |

**Table 4.5.** *Severity levels of the classes of a system (Flaus 2019)*

| Likelihood | Designation |
|---|---|
| L1 | Very low |
| L2 | Low |
| L3 | Medium |
| L4 | High |

**Table 4.6.** *The ANSSI likelihood scale*

| Level | Designation | Significance |
|---|---|---|
| L1 | Very unlikely | Occurs less than once every 100 years |
| L2 | Possible | Occurs less than once every 10 years |
| L3 | Occasional | Occurs less than once a year |
| L4 | Frequent | Occurs more than once a year |

**Table 4.7.** *Simple likelihood scale (after Flaus (2019))*

| Severity/Likelihood | L1 | L2 | L3 | L4 |
|---|---|---|---|---|
| Catastrophic | 2 | 2 | 3 | 3 |
| Major | 2 | 2 | 2 | 3 |
| Moderate | 1 | 2 | 2 | 2 |
| Minor | 1 | 1 | 2 | 2 |
| Insignificant | 1 | 1 | 1 | 1 |

**Table 4.8.** *Cybersecurity risk matrix. The color code for each cell qualitatively indicates the intensity of the risk: green = low; yellow = moderate; orange = large; and red = critical. For a color version of this table, see www.iste.co.uk/laurent/processsafety4.0.zip*

## 4.5. Coordinating risk analysis methods

Coordinating the application of process safety risk analysis methods and cybersecurity risk analysis methods is essential. Figure 4.4 compares the organigrams for the steps used in both methods and reveals interactions 1, 2 and 3 between the processes used in both analyses.

Some essential interactions must take place between the respective stakeholders:

– Area 1: a physical analysis of process safety identifies scenarios that are harmful to people, goods and the environment. These results are sent to the entity carrying out the cybersecurity analysis so that they can be integrated as hazardous events in the cybersecurity risk analysis.

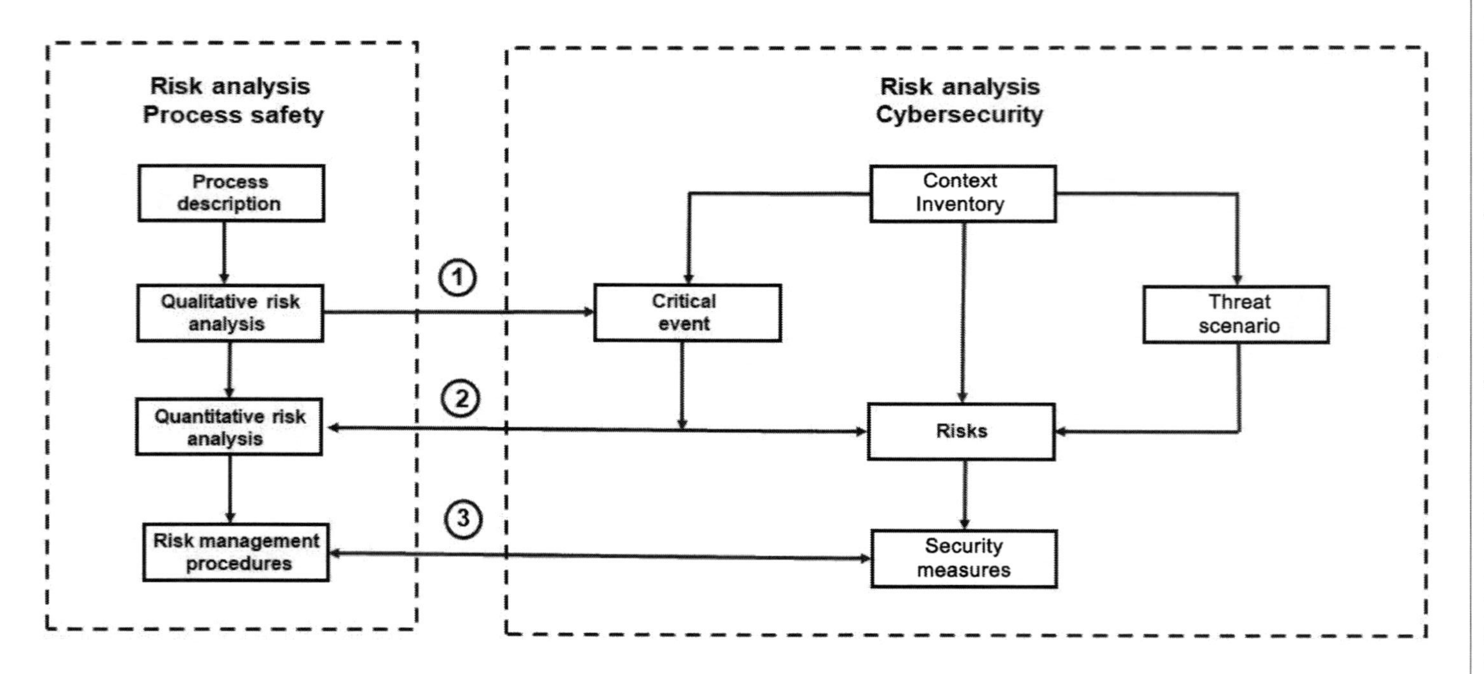


**Figure 4.4.** *Comparison and interactions between organigrams for risk analysis methods (Masse et al. 2018)*

– Area 2: during the step where there is a detailed risk analysis for each of the two methods, the allocation and specification of the safety integrity levels (SILs) of safety instrumented systems (SISs) must feed into the cybersecurity analysis as the SIS may be subject to cyber-risks. The converse is applicable for security levels (SLs) of cybersecurity. Information exchange implies that the likelihood of a cyberattack is integrated, with its random and malicious causes, into the bow-tie model.

– Area 3: in terms of risk control and safety measures, reciprocal sharing of information requires defining and specifying the demands of the measures required to arrive at an acceptable level of risk.

## 4.6. Reconciling process safety and cybersecurity methods

The functional safety standard IEC 61511, version 2018, from the IEC now mandates a risk assessment concerning the safety of SISs. The ISA has published a technical report (ISA-TR84.00.09-2017), which documents a process to assess cybersecurity risks to SIS. The IEC 62443 standard is applicable to monitoring systems in industrial installations, which contain systems with automation, SCADA supervision, network steering and IIoT. It looks at controlling all aspects of industrial cybersecurity from the point of view of all the stakeholders. This cybersecurity risk assessment must make it possible to identify the most serious consequences to health, safety, security and the environment of the system and to list all hazardous scenarios where the trigger event and all safety barriers can be attacked or hacked. In terms of reconciling the methods, the different risk analysis methods can be used to determine the impact of cybersecurity threats and vulnerabilities on process safety. On the other hand, these methods must also be adapted to take into account the cybersecurity risks related to the integrity of the safety barriers selected for each specific risk scenario.

### 4.6.1. *Preliminary risk analysis and preliminary cyber-risk analysis*

#### 4.6.1.1. *Preliminary risk analysis*

The simple preliminary risk analysis (PRA) method consists of identifying the various hazardous elements present in the industrial process being studied and examining each of these to see how they could lead to a more or less serious accident situation following an event that would initiate a potentially hazardous situation. This analysis allows us to deduce all means and all corrective actions that would make it possible to control or eliminate these hazardous situations and potential accidents. The preliminary risk analysis method is composed of several steps:

– listing the disadvantages, hazards and links related to the different products being used (reactants, products, intermediaries, etc.) and to their inherent properties;

– identifying the risks related to the chosen processes (principal and secondary chemical reactions, etc.) and to the equipment and technologies used;

– defining the different arrangements adopted in the organization and the installation procedures, as well as the means of intervention in the case of an accident;

– bringing in a response that is appropriate for the different risks identified with the aim of controlling the residual risk in the unit by maintaining it at an acceptable level.

The information resulting from this preliminary risk analysis is entered into specific tables. At present, there is no consensus on the formats and degree of resolution to use for these tables that could lead to a specific codified representation. The detailed steps for carrying out a PRA are described through examples in the texts (UICh 1980; Laurent 2011).

#### 4.6.1.2. *Preliminary cybersecurity risk analysis (Cyber PRA)*

The Cyber PRA method is an extension of the PRA method that takes into account cybersecurity problems by adding the information technology causes for a hazardous attack that could disrupt a scenario, leading to failure of safety barriers and requiring a correction of the assessment of risk control measures.

Flaus (2019) suggests the following steps for carrying out the Cyber PRA method:

– preparing for the analysis (context, perimeter, Return of EXperience (REX), metrics, etc.);

– describing or modeling the installation, broken down into systems that make it possible to carry out the functions, especially information processing systems (BPCS-SIS);

– analyzing hazardous events that could induce a loss of control over the process and the consequences of these;

– looking for initiating cyber-events;

– analyzing the consequences of each hazardous event;

– listing the existing cyber-attackable barriers;

– assessing the severity and likelihood (concatenated matrix);

– improving the risk matrix by adding barriers, whether physical or instrumented, with an SL.

### 4.6.2. HAZOP, CHAZOP and Cyber HAZOP methods

An overview of the principle behind these three methods illustrates the advantages of their respective complementary strengths when it comes to reconciling methods, although they are often considered contrasting methods.

#### 4.6.2.1. *HAZOP method*

The principle behind the HAZOP methods consists of first describing the operation of the different phases in a process in detail. This is done by breaking it down into a series of elementary operations using detailed plans of the installation. Next, the various possible errors and deviations from the installation's parameters are listed using a list of keywords or guide-words. Thus, each keyword describes a kind of deviation, whose causes are identified and whose consequences are studied. The systematic reference to a keyword expresses the potential loss of function in the elementary operation, the subsystem or the system. The deductive analysis that is applied makes it possible to list the failures, while the inductive analysis leads to a list of effects (Laurent 2011).

This entire analysis leads to the establishment of grids that indicate the possible causes for the deviations, their consequences and the actions or technical modifications required to ensure the proper functioning and/or safety and security of the system. It must be noted that some deviations could impact the production without systematically generating harmful or hazardous consequences with respect to the functioning of the installation.

Finally, potentially hazardous deviations are listed in a hierarchy to determine the actions to take.

To summarize, the HAZOP method is based on the following systematic principles:

– choosing an operating parameter;

– generating a deviation from this parameter using a list of keywords;

– finding the possible causes for the deviation being studied;

– determining any consequences that may be associated with this deviation;

– in the case of hazards, verifying the presence or existence of correction, prevention, protection or mitigation systems. In the absence of correction systems, it is necessary to define them all.

#### 4.6.2.2. *CHAZOP method*

The CHAZOP method analyzes the effects of potential deviations in the industrial command control system on changes in the installation. The CHAZOP method must be considered here as an extension of the HAZOP method. While the HAZOP method limits its analysis to the high, low or zero deviation of a failure in a control loop, the CHAZOP method extends this to search for common causes for this failure. The basic approach in the CHAZOP method is identical to that of the HAZOP method (control list, keywords). The analysis covers both the hardware and software aspects of the computerized control system. Typical keywords are as follows: no, more, part of, other than, early, late, before, after, etc. (Table 4.9). Variations in these keywords are implicitly included. These keywords can be applied to the following domains:

– communications (data signals);

– digital hardware (processor, I/O);

– mechanical components, chiefly the components of origin and the destination in the command loop, for example, sensors and locking valves. The detailed interactions between these components can only be considered if it is essential to understand the modes of failure to assess the deviations with the design intention or if it is essential for interconnections.

| **Subsystem communication** | |
|---|---|
| Keyword | Deviations |
| no | Signal (zero read, full scale read) |
| more | More current, erratic signal |
| part of | Incomplete signal |
| other than | Excessive noise, corrupt signal |
| early | Signal generated too early (timer problems) |
| late | Signal generated too late |
| before/after | Incorrect signal sequence |
| Subsystem Digital Hardware | |
| Keyword | Deviations |
| no | I/O failure |
| more | Multiple failure (control card, processor rack, processor) |
| part of | Partial failure of card, failure of counters |
| other than | Abnormal temperature, dust |
| Subsystem Software | |
| Keywords | Deviations |
| no | Program corruption |
| more | Memory overflow |
| part of | Addressing errors/data failure |
| other than | Endless loops, data validation problems, operator override |
| early/late | Timeout failure, sequence control problems, sequence interpretation error |

**Table 4.9.** *Partial example of the use of keywords and deviations in the CHAZOP method*[1]

It is useful to draw up lists of specific keywords for each analysis:

– loop diagrams of the SIS, schematic diagrams or organigrams;

– electronic circuit diagrams, or, if necessary, cause and effect diagrams for instruments;

1 Avilable at: http//www.uobabylon.edu.iq/eprints/publication_7_13123_6247.pdf.

– PID schematic diagrams to identify the impact of system deviations on the process.

#### 4.6.2.3. *Cyber HAZOP method*

Flaus (2019) proposed an approach, called the Cyber HAZOP, which is an extension of the HAZOP method, and which explicitly handles aspects related to cybersecurity. The sequence of steps is similar to that used in the HAZOP study:

– Step 1: defining the system, its subsections and its nodes. Selecting the limits of the system, dividing it into manageable subsystems and identifying the nodes to consider.

– Step 2: defining the problems of concern and specifying the problems of concern that we will be analyzing in terms of hygiene, safety, security and environment.

– Step 3: applying the deviations to each node and developing scenarios for significant threats (causes/hazards).

– Step 4: examining the consequences and identifying all significant implications for each deviation without, at first, considering existing safeguards.

– Step 5: examining the causes and identifying the vulnerabilities or potential causes of deviations.

– Step 6: calculating the non-attenuated risk and evaluating the probability and impact of this deviation.

– Step 7: identifying the protection measures and determining the most solid protective measures against each consequence.

– Step 8: calculating the residual risk and re-evaluating the probability and impact of the deviation with respect to existing protection measures.

– Step 9: identifying and recommending additional protection measures, then specifying other protection measures to bring down the risk to an acceptable (tolerable) level.

– Step 10: recording, establishing the report, finalizing recommendations for safeguards in order of priority.

In terms of cybersecurity, this procedure is modified during the step of identifying means of prevention, mitigation and protection. In particular, if a safety barrier chosen for a deviation is attackable, a cybersecurity risk assessment must be carried out in addition to the conventional process safety analysis. In this case, the assessment of the consequences must guide the choice of safety measures. A non-attackable, physical barrier is recommended to prevent large or catastrophic consequences. For other consequence levels, it is possible to further reinforce the required SL according to the nature of the violations.

### 4.6.3. *Bow-tie graph and cyber bow-tie*

#### 4.6.3.1. *Bow-tie graph*

The principle behind the bow-tie graph has already been reviewed in section 4.1. This graph can conventionally be used to specifically analyze the cybersecurity of an industrial control system. Figure 4.5 presents an example of a bow-tie model that describes a malware attack on a historian server, along with its causes and consequences. This global presentation has been used, as part of intra-organizational training, to initiate operators qualified to work in a chemical process unit.

#### 4.6.3.2. *Cyber bow-tie*

Flaus (2019) has indicated that it is also possible to consider causes arising from cyberattacks within a conventional bow-tie graph. These causes are linked either to a loss of integrity in the software or in the command system data or to a lack of SIS availability. For example, Figure 4.6 thus illustrates the graph event, initially physically disrupting the process and situated upstream of the central critical event. The initial causes acting on this $Ein6$ event are called the "initiator cyber-events", identified here by the label $CyberEIn$.

### 4.6.4. *LOPA and Cyber LOPA methods*

#### 4.6.4.1. *LOPA method*

LOPA (Layer of Protection Analysis) is an integrated probabilistic risk assessment method. This method proposes assessing risk control measures based on the "barrier" approach by analyzing the contribution of different successive protection layers. It is one of the methods cited in the IEC 61511 standard to estimate SILs required for instrumented safety functions. The LOPA method can be applied either in a semi-quantified, discrete-probabilistic manner, or in a completely probabilistic manner. The implementation of this method is structured into a six-step process:

– Step 1: "identifying the consequences of the scenarios" aims to list the scenarios that will have a significant impact in terms of severity and consequences. This step makes it possible to limit the number of scenarios to study and thus reduce the duration of the LOPA study.

– Step 2: "selection and developing scenarios" uses the inventory created in the previous step and consists of using conventional, qualitative risk assessment tools (PRA, HAZOP, trees, etc.) to clearly define the steps in the scenarios and represent each scenario through a bow-tie diagram.

– Step 3: "identifying the set of initiating events and evaluating their frequency" proposes using the detailed analysis of the chosen scenarios to identify the set of initiating events which, if they occur, can lead to the central critical event. The frequencies of these EIs is estimated using occurrence values derived from the return of experience and data from literature or expert opinions.

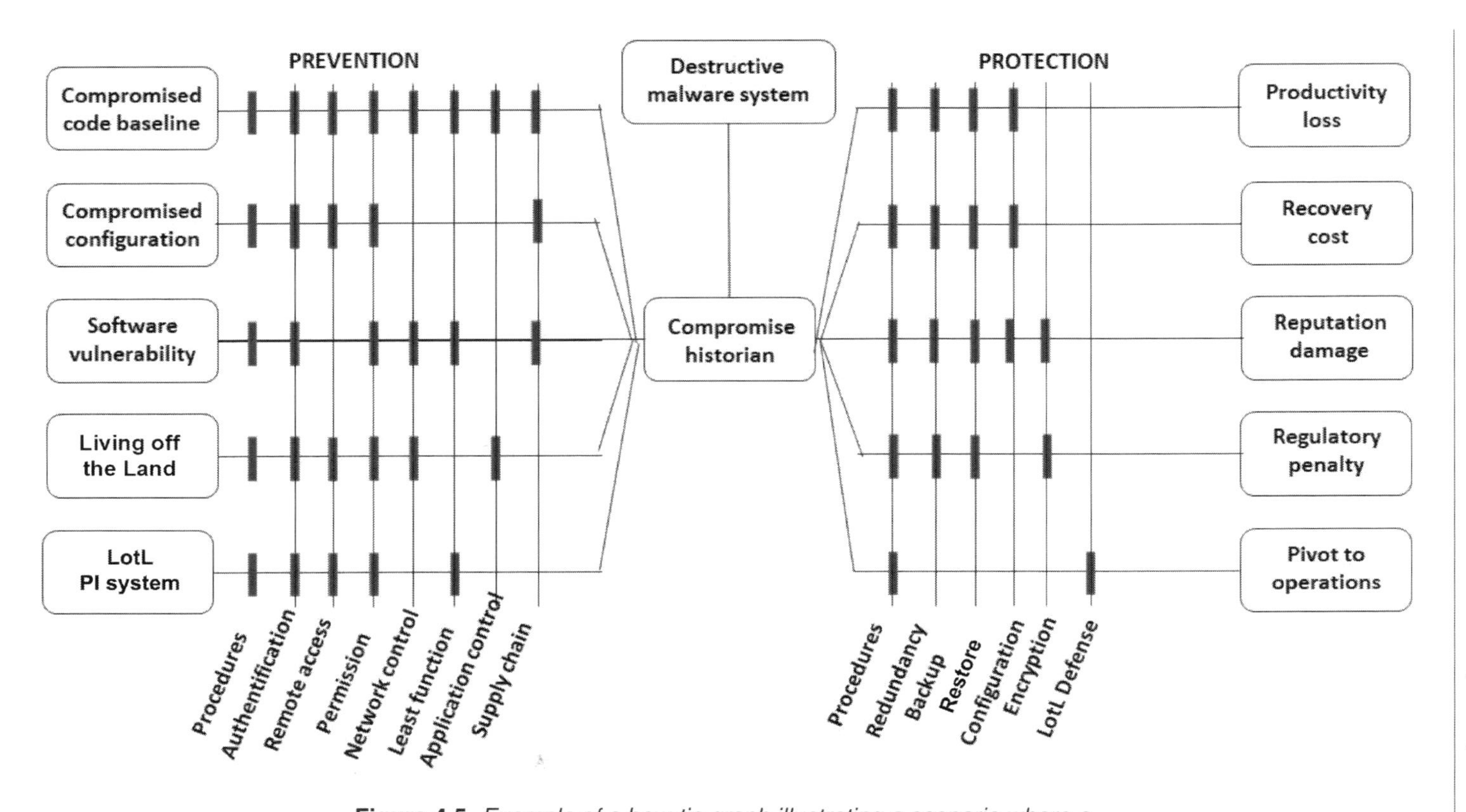


**Figure 4.5.** *Example of a bow-tie graph illustrating a scenario where a historian server is attacked by malware*

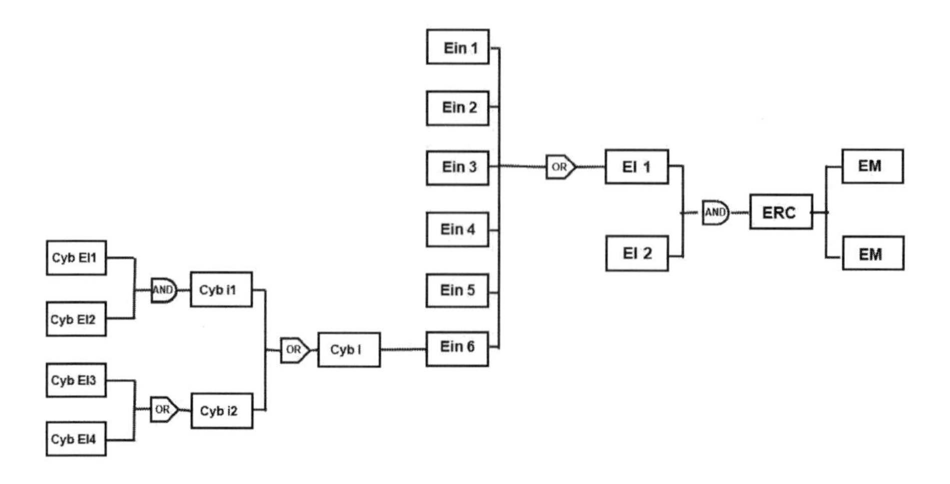


**Figure 4.6.** *Schema showing the principle of a cyber bow-tie graph coupling cybersecurity causes and process security causes*

– Step 4: "identifying the functions and safety barriers and assessing their failure probabilities" makes it possible to identify, for each scenario, measures that can prevent the unfolding of the scenario leading to the central critical event by considering the qualification criteria for the devices chosen, especially the IPL (Independent Protection Layer). A failure probability value is then associated with each safety barrier, noting that this LOPA method step explicitly references the SIL used in the IEC 61511 – 3 standard. Stack (2009) highlighted the difficulty in identifying the true independence of the protection layers and indicates an assessment procedure in the case where the barriers are not independent. It must also be noted that this step makes it possible to consider all the barriers, whether technical, human or organizational (Gowland 2006). Baybutt (2002) has proposed the LOPA-HP method to improve the estimation of failure probabilities of human barriers.

– Step 5: "estimating the probability of the occurrence of the central critical event (CCE)", which consists of bringing together information from the previous steps. This step makes it possible to calculate the probability of occurrence of the CCE by combining the probabilities of occurrence of the initiating event (EI) and the failure of the protection and attenuation layers concerned.

– Step 6: "assessing the risk with respect to risk control criteria" involves ensuring that the risk is well controlled with respect to pre-defined criteria, for example, a severity-frequency criticality grid, or a global criterion for the cumulative risk for a process or for a site.

#### 4.6.4.2. *Cyber LOPA method*

Tantawy et al. (2019, 2020a, 2020b) proposed an extension of the LOPA method, called Cyber LOPA, to integrate the potentiality of cyberattacks into the risk analysis of an industrial process involving a cyber-physical information and industrial control system. Since the safety requirements for such a system must be defined as early as possible in the design and pre-sizing step for the process, the authors simultaneously consider that:

– first, cybersecurity parameters impact process safety;

– second, process safety has a reciprocal impact on elements of cybersecurity.

Figure 4.7 illustrates this interdependence in the proposed cybermodel.

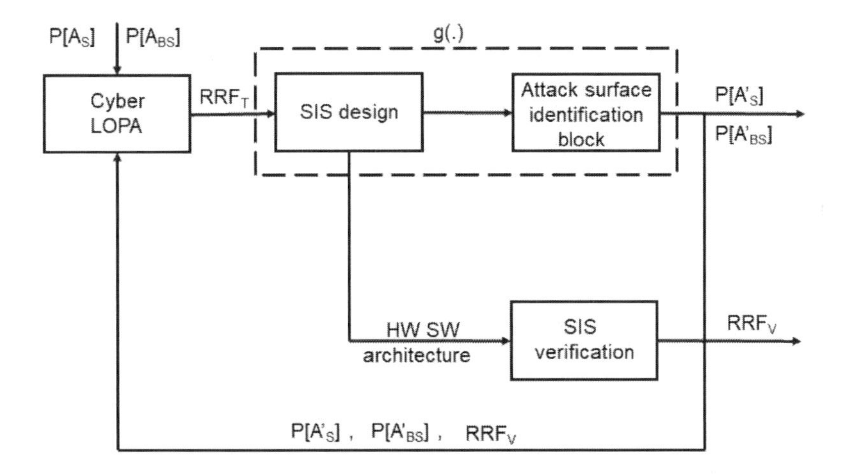


**Figure 4.7.** *Model showing the interdependence of aspects of safety and cybersecurity (Tantawy et al. 2020a)*

The computation algorithm is initiated by the reasonably acceptable values chosen for the direct probability of failure of the SIS, $P[A_S]$, and the risk reduction target factor $RRF_T$. The corresponding value of the probability of failure, $P[A_{BS}]$, of the BPCS SIS pivot system is calculated from the Cyber LOPA system. By using $RRF_T$, the design process for the SIS is launched and the hardware and software architectures for the entire system are produced. The attack surface is then identified for the architecture produced, and the probabilities of the cyberattack, $P[A'_S]$ and $P[A'_{BS}]$, are estimated. In addition, the $RRF_V$ resultant of the design and verification processes for the SIS is calculated. This resultant $RRF_V$ is not necessarily the same as the desired input $RRF_T$. It is generally greater for a successful SIS design cycle. The square in dashed lines illustrates the fact that the cyberprobabilities also depend on the probability of physical failure via the combined blocks of "SIS design" and "identification of the attack surface", with this functional dependency denoted by the vector function g(.). If $P[A'_S]$ and $P[A'_{BS}]$ generate a new $RRT' \leq RRT_V$ value via the Cyber LOPA model, the design is then finished. In the contrary case, a new iteration is necessary.

Tantawy et al. (2020b) have reported the case study of a continuous-stirred tank reactor (CSTR) where a first-order chemical reaction, with respect to the reagent, is implemented. The CSTR's functioning occurs through a BPCS industrial control system in accordance with the NIST 800-82 standard. These authors have quantitatively demonstrated the advantage of the Cyber LOPA method, taking into account the cyberattack failures and by quantifying the differences in estimation, with respect to the application of the LOPA method. Cormier and Ng (2020) have also highlighted the important of integrating cybervulnerability analyses into the LOPA method.

Iddir (2021) has noted that the LOPA method could be associated with other quantified risk analysis method. For example, the initial "cause versus consequence" approach of the LOPA method is comparable to an event tree that would focus only on the sequence of the tree associated with the failure of the barriers. Similarly, during the analysis of complex accidents resulting from multiple sequences that involve multiple scenarios, the LOPA method is applied to the unit-wise handling of each scenario and then their final aggregation is further refined by a bow-tie approach.

### 4.6.5. *The integrated, simultaneous ATBT method*

Abdo (2017), Abdo et al. (2018) and Masse et al. (2018) proposed the ATBT (Attack Tree Bow-Tie) method, combining bow-tie graphs and attack trees, making it possible to identify malicious causes for attacks in the scenarios described in the conventional process risk analyses. This representation makes it possible to bring together cybersecurity risks and risks related to the industrial process safety.

The cybersecurity risk $R_{cyber}$ is defined by the relation:

$$R_{cyber} = [tv_j; P_{(tv)_j}; X_{(tv)_j}]; \text{ with } j = 1, 2...M$$

where $tv_j$ represents the attack $t$ that exploits the vulnerability $v$, $P_{(tv)_j}$ is the probability that the threat $t$ will exploit the vulnerability $v$, $X_{(tv)_j}$ is the severity of the consequences if $t$ exploits the vulnerability $v$ and $M$ is the number of possible attacks.

The set of attack scenarios are illustrated by attack trees representing the different vulnerabilities exploited by an attacker to produce the undesirable effect on the physical system of the process.

The process safety risk $R_{safety}$ is defined by the relation:

$$R_{safety} = [S_{(e)_i}; P_{(e)_i}; X_{(e)_i}]; \text{ with } i = 1, 2...N$$

where $S_{(e)_i}$ is the scenario representing the undesirable event, $e$, its causes and its consequences; $P_{(e)_i}$ is the probability of occurrence of $S_{(e)_i}$, $X_{(e)_i}$ is the severity of the consequences of the scenario $S_{(e)_i}$ and $N$ is the possible number of scenarios or the number of undesirable events causing damages.

The set of scenarios related to process safety is integrated into the bow-tie graphs in the conventional risk analysis framework.

The global risk, $R$, which simultaneously expresses cybersecurity and process safety, is defined in terms of a triplet through the relation:

$$R = [S_{(tv,e)_i}; P_{(se,sa)_i}; X_{(tv,e)_i}]; \text{ with } i = 1, 2...N$$

where $S_{(tv,e)_i}$ is the description of the scenario with the undesirable event, $e$, which could result from safety incidents and/or violations of cybersecurity, where $tv$ is the threat that exploits the vulnerabilities; $P_{(se,sa)_i}$ is the probability of occurrence of the scenario, where $se$ and $sa$ are, respectively, the probabilities related to cybersecurity and safety; $X_{(tv,e)_i}$ is the severity of the consequences of the scenario; and $N$ is the number of possible scenarios or individual events that could result in damage.

The methodology of simultaneously analyzing cybersecurity and safety risks is composed of three main steps:

– *Identifying risk scenarios*: the method combines the bow-tie graph and attack tree to identify the causes and consequences associated with cybersecurity and the safety of the undesirable event being considered. This makes it possible to identify and take into account all incidents and threats that could lead to the same damage-causing undesirable event.

– *Assessing the probability*: since the bow-tie graph and attack tree both offer the respective probability assessment for the risk scenarios related to process safety and cybersecurity, combining them offers the same option for risk scenarios associated with cybersecurity and safety. Particular attention must be paid to this assessment. Since the RS for cybersecurity and safety are different in nature, the probability of the events related to process safety will appear to be very small compared to the probability of cybersecurity events. This is why we must consider different scales to measure the probability.

– *Evaluating the severity of the consequences*: this step aims to quantify losses in terms of human life, environmental damage and assets in the system in the case of the undesirable event occurring, by considering that the severity of an individual scenario is the same, regardless of whether the cause is related to safety or cybersecurity.

Abdo (2017) has summarized the procedure for simultaneously analyzing cybersecurity and process safety risks in the organigram shown in Figure 4.8.

After constructing the bow-tie graph for each scenario (process safety), the attack graph is established (cybersecurity), and then attached and combined to the bow-tie graph. The contents of the box in dashed lines illustrate the steps in assessing the probability of the combined scenario.

Abdo et al. (2018) applied this procedure to the operation of a discontinuous-stirred tank reactor (batch), carrying out a highly exothermic reaction under pressure, susceptible to producing a thermal runaway reaction. The cooling equipment in the process was controlled by a programmable logic controller, supervised by an industrial SCADA control system. The information that was collected was accessible to the operators through a wireless remote. The installation could also be operated via Internet through a tablet or smartphone.

## 4.7. Concatenation of matrices

The distinction between the different areas of risk is not an easy one and chiefly depends on the definitions given to the different probability and severity levels that make up the criticality matrix. It is proposed that we study the concatenation principle of a conventional process safety matrix and a cybersecurity matrix for an industrial installation.

Let us consider, for instance, the probability–severity matrix for process safety risks. Probability here is defined by five levels, denoted by the letters E to A, in increasing order of probability. The severity is expressed by five levels in increasing order: insignificant, minor, moderate, major and catastrophic. Each cell in the matrix contains a number along a scale from 1 to 4 in increasing order of risk (Table 4.10a).

Let us look at the likelihood-severity cybersecurity matrix for the corresponding installation. The qualitative likelihood scale is indicated by 4 levels, labeled $L1$ to $L4$, expressing, respectively, a very low likelihood, low likelihood, moderate likelihood and high likelihood. The severity scale is chosen here, simply through homogeneity, identical to the scale used for the process matrix. Each cell in the matrix contains a number from 1 to 3 in increasing order of risk, corresponding to the system class (Table 4.10b).

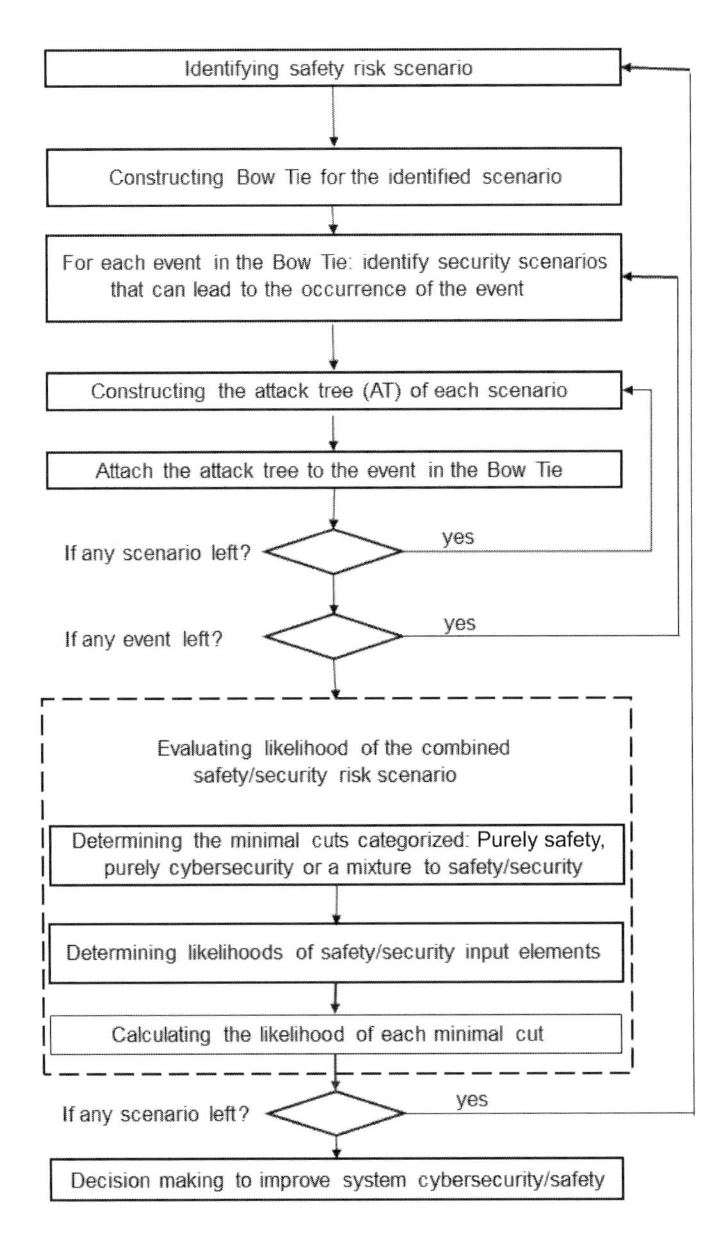


**Figure 4.8.** *Organigram for the simultaneous ATBT procedure for cybersecurity and process safety risks (Abdo 2017)*

| Gravity/Probability | E | D | C | B | A |
|---|---|---|---|---|---|
| Catastrophic | 2 | 3 | 4 | 4 | 4 |
| Major | 2 | 2 | 3 | 3 | 4 |
| Moderate | 2 | 2 | 2 | 2 | 3 |
| Minor | 1 | 1 | 1 | 2 | 2 |
| Insignificant | 1 | 1 | 1 | 1 | 1 |

a)

| Severity/Likelihood | L1 | L2 | L3 | L4 |
|---|---|---|---|---|
| Catastrophic | 2 | 2 | 3 | 3 |
| Major | 2 | 2 | 2 | 3 |
| Moderate | 1 | 2 | 2 | 2 |
| Minor | 1 | 1 | 2 | 2 |
| Insignificant | 1 | 1 | 1 | 1 |

b)

**Table 4.10.** *a) Process safety risk matrix; b) cybersecurity risk matrix: green = low; yellow = moderate; orange = large; and red = critical. For a color version of this table, see www.iste.co.uk/laurent/processsafety4.0.zip*

The concatenation of the two matrices presented in Table 4.11 makes it possible to obtain what is called an "extended" matrix, which expresses the overall level of risk for the installation, thus taking into account process safety and cybersecurity.

The concatenation must be carried out carefully and expertly. Indeed, a scenario with a chain of events leading to the CCE may be generated, solely through a physical failure in the process, or solely through a cyberattack, or through a combination of the two. In the first case, the assessment is based on the probability (P) of the conventional process safety matrix. In the second case, the assessment is carried out using the likelihood level (L) of the cybersecurity matrix. In the case where the scenario is has two origins, the concatenated matrix is usable with the probability (P)–likelihood (L) couple. However, it must be recalled that for a given consequence, the concatenation tries to create a quantitative and qualitative matrix.

## 4.8. Reasoned use of risk matrices

Risk matrices are widely used and sometimes even mandatory in an industrial context. It makes it possible to plot a scenario for a system event in a probability–severity plane. The major problem in using risk matrices is considering that they truly measure risk. The ease of reading these matrices is seen as an advantage for all stakeholders, including the general public, as there is no specific knowledge needed to understand them. On the other hand, the result that is displayed is not systematically pertinent, at least not statistically. Consequently, using these matrices can create a false sense of security within the organization being assessed. The risk assessment scores carried out by different teams of international experts on the same process have shown that significant deviations were obtained (see Fabbri and Contini (2009); Markert et al. (2000); Allford and Wood (2021)). The bias induced by the experts is likely to play a major role in the use of these methods and data during the assessment. Indeed, the availability of the data, its relevance and the uncertainty in this data are important factors to consider. For example, the REX for

| | Probability – Likelihood | | | | | | | | | | | | | | | | | | | |
|---|---|---|---|---|---|---|---|---|---|---|---|---|---|---|---|---|---|---|---|---|
| | L=1 | | | | | L=2 | | | | | L=3 | | | | | L=4 | | | | |
| | Probability | | | | | Probability | | | | | Probability | | | | | Probability | | | | |
| **Severity** | **E** | **D** | **C** | **B** | **A** | **E** | **D** | **C** | **B** | **A** | **E** | **D** | **C** | **B** | **A** | **E** | **D** | **C** | **B** | **A** |
| Catastrophic | | | | | | | | | | | | | | | | | | | | |
| Major | | | | | | | | | | | | | | | | | | | | |
| Moderate | | | | | | | | | | | | | | | | | | | | |
| Minor | | | | | | | | | | | | | | | | | | | | |
| Insignificant | | | | | | | | | | | | | | | | | | | | |

**Table 4.11.** *Extended concatenated process safety risk matrix.*
*The color code for each cell qualitatively indicates the intensity of the risk: green = low; yellow = moderate; orange = large; and red = critical. For a color version of this table, see www.iste.co.uk/laurent/processsafety4.0.zip*

the use of nuclear centers of the same type, oil refineries that are similar, offshore platforms with identical models, all make it possible to use a lot of acceptable data. Nonetheless, there is no statistically exploitable data for chemical industry processes that are very different in nature, and, a fortiori, the new technology used in Industry 4.0.

Cox (2008), Quintino (2011) and Petrusich and Schwarz (2017) have highlighted, in addition to the large variety of estimations from experts, several other, inherently limiting problems in quantitative risk assessment matrices:

1) risk matrices based on hypotheses, experience and generic data are sub-optimal;

2) a prerequisite also implies that the maintenance of the system, unit or plan be effective;

3) another hypothesis also involves that the system, unit or plant be used by competent and available operators;

4) always verifying the type of content in the matrix, either a rough process matrix, or a matrix containing all the added safety measures, ensures that the maximum acceptable risk level is not exceeded;

5) their low resolution: typical risk matrices can accurately and unambiguously compare only a small fraction (less than 10%, for instance) of the pairs of hazards selected at random; they could attribute identical scores to risks that are quantitatively very different (range compression concept);

6) computation errors: risk matrices could erroneously attribute higher qualitative values to risks that are quantitatively smaller; for risks whose frequency and severity are negatively correlated, they could be "worse than useless", leading to decisions that are poorer than they are uncertain;

7) sub-optimal resource allocation: efficient resource allocation to risk-reduction countermeasures cannot be based on the categories given by the risk matrices;

8) the ambiguity in inputs and outputs: for example, the metric of the severity scale cannot be objectively chosen for uncertain consequences. As a result, the input metrics (frequency and severity) and the resultant output risk score require subjective interpretation. Consequently, different users could propose contradictory assessment of a quantitatively identical risk.

REMARK.– *To improve the evaluation, it is possible to overcome some of these difficulties, for example, by using the Monte Carlo simulation method and/or Bayesian networks.*

*The Monte Carlo simulation is a statistical analysis that consists of computing a numerical value by using random processes within the parameters of a defined distribution. This probabilistic technique makes it possible to determine the expected values for risks based on data from previous processes. The Monte Carlo simulation generates a large number of random variables and provides the distribution of the results. The two parameters used to analyze the risks are, conventionally, occurrence and risk impact. Each iteration of the simulation generated the product of the occurrence and impact of the risk. The quality of the results of such a simulation resides in the choice of the type of distributions for both these parameters for each of the risks analyzed. These distributions must be specified in the model before the simulation is run. One of the most widely used distributions for modeling is the triangular distribution. This distribution has the advantage of only requiring the estimation of three defined scores by the participating experts: minimum, most probable and maximum. This estimation also makes it possible to take into account the uncertainty of the experts.*

*Bayesian networks are oriented acyclic graphs, where the variables are represented by nodes and relations of dependency or correlations between the variables are represented by directional arcs. Each variable is represented by a table of probabilities, which are determined by using Bayes' theorem. This theorem offers adaptability and flexibility, which allow the user to revise and change the estimations and predictions if new, pertinent data becomes available. The chief advantage that Bayesian networks offer is that partial or incomplete data can be used and these networks are easy to adapt to a specific environment through machine learning. They also make it possible to study causal relations and the direct influence of one variable on another. They combine expert estimations and statistical data to better evaluate causality, making it possible to associate all available data sources, whether subjective or objective. These networks carry out cause and effect reasoning in a transparent and documented manner, while also allowing the acquisition, representation and use of knowledge. The limitation of using Bayesian networks lies in the high complexity of integrating it into a framework based only on expert opinions, into heavy graphs and computational algorithms with complex networks, and in the difficulty of working with continuous variables.*

# 5

# Examples: Safety 4.0 and Processes

This chapter aims to showcase different examples that illustrate the position and influence of Safety 4.0 in the design, implementation, use and revamping of future processes. The ex nihilo design of new processes, or even of products, is virtually non-existent in published literature due to well-known restrictions related to confidentiality and the development of innovation. The majority of these examples are a result of the improved use of existing processes and the need to revamp certain processes in order to improve industrial, environmental and social conditions. The important position of intensification processes will be studied in Chapter 6 when discussing the specific criterion of safety.

## 5.1. Distillation column control

This simple example demonstrates the influence of models in the automatic control of a simple distillation column. Quality indicator measurements, such as the composition of key components in the distillate or the bottom products of the distillation column, must systematically be available online. In reality, these online measurements are not frequently used or are still sometimes carried out in a laboratory at an interval of a few hours. However, it is necessary to monitor these indicators of the composition of key components in an online database, with a frequency of the order of 1 minute, for example. Inferential models are used in order to provide information at this frequency. The concept of an inferential model consists of calculating a property of the streams based on easily accessible process measurements, such as temperature and pressure, to avoid direct acquisitions that are too long or costly. The accuracy of these inferential models, however, must be periodically reviewed and verified. These inferential models may be used to access information on the operator, cascading toward the controllers of the base layer process or output variables of multivariable controllers. The performance of an advanced process command (APC) application depends enormously on the quality of these estimation models.

For a binary distillation column, the client did not find the performance of the inferential model implemented to estimate the quality of the bottom of the column to be satisfactory. It was thus necessary to design a new estimation model. It was seen that the estimation model used was redundant in terms of input temperatures and also insufficient in terms of measurements for the material balance. The combination of the correlation between the inputs and the choice of structure for a dynamic model led to identification results consisting of the temperature profile of the foot of the column, the respective temperature profiles of the two parts of the enrichment section and the pressure drop in the column. A symptom of over-adjustment was observed via the sign of gain in the model between the quality at the foot and the temperature after a part of the column section, which changes with time. This indicates that the overall identification result is not reliable.

A new static model was thus identified, taking into consideration only a single temperature measurement (from the bottom stream of the column) and an indicator of the material balance in the column (reflux/feed relationship). With the more satisfactory solution from this model and the gains from a more statistically significant model, the new estimation model leads to better estimations, with respect to the earlier formulation. The new estimation model is capable, in particular, of better predicting the rare measurements on the bottom of the model, especially at high or low concentration values[1].

## 5.2. Attempt to classify the applications of a digital twin in the field of Safety 4.0

The definition for a digital twin was presented in Chapter 1 (section 1.3.2.7) as a communication and interconnection technology used in Industry 4.0. To recall this, a digital twin is a virtual clone or a replica of a physical system or a process. It systematically involves the existence of a pair made up of the digital model and the object of which it is a copy. The concerned objects could be a product, a machine, a production line, a process or a supply chain.

The primary purpose of a digital twin is to simplify access to information that are integrated into the virtual representation of the production site from which the data are taken. The digital twin also allows teams to carry out virtual visits to the production sites, thus saving the teams having to visit the site physically to carry out a study for an intervention, perhaps. The virtual twin also facilitates remote command and/or surveillance operations via sensors that track the parameters. Finally, digital representation can take into account the operational dynamics of

1 Complementary details may be found at: https://ipcos.com/cases/revamp-of-inferential-models-for-distallation-column/ and https://ipcos.com/cases/strategies-for-successful-implementation-of-indus- try-4-0-solutions/.

equipment or of a unit. By integrating artificial intelligence, machine learning and data analysis, the digital twin generates updated simulation models by automatically adapting to changes.

### 5.2.1. *Potential of a digital twin for Safety 4.0*

According to Agnusdei et al. (2021), to understand how a digital twin contributes to the field of Safety 4.0, the following observations must be considered:

– The capacity for a digital twin to dynamically update data from the physical world to the digital world, and vice versa, could help develop dynamic risk models. Consequently, the risk assessment process becomes more reliable as real information can be provided about the knowledge of equipment or a workplace in order to evaluate the operating states of a hazardous system.

– The presence of data-processing tools integrated into the virtual twin can support more efficient predictive analysis of scenarios for problems involving complex hazards. Based on the type of data processing tool present in the specific digital twin application, more complex abilities could be used to widen the scope of analysis of the safety problem.

– Acquisitions of two-way flows (from the physical world to the digital world and vice versa), generally carried out by a digital twin, could contribute to the development of early warning tools allowing the implementation of proactive actions in complex worksites.

### 5.2.2. *Proposal for a classification framework*

Based on current applications of digital twins, Agnusdei et al. (2021) proposed a framework to evaluate the global capabilities of a specific twin applied to industrial safety. The goal is to propose a tool to promote the development of new applications for digital twins that will allow more efficient design, assessment and management of safety in industrial systems. The framework is based on three main criteria related to the principal characteristics of digital twin applications. These are data acquisition, data processing and safety criteria:

1) The "data acquisition" criterion refers to the manner in which data are acquired from the physical world. Based on the tools and technologies adapted in current applications of a digital twin, this criterion has been divided into three sub-categories representing different tools, characterized by increasing levels of complexity and reliability:

- random data;
- historical data;
- real-time data from physical sensors.

The first two categories refer to non-simultaneous acquisition of real data from a process or a product. They are generally stored in a data reference system and can then be used in the digital world. Real-time data acquired by physical sensors are the top of the data acquisition intensity scale and thus represent the most useful acquisition method to support a digital twin. The data acquisition criterion can thus be represented along an axis by an incremental scale, which varies from a minimum to a maximum (i.e. random data to real data) based on the intensity of support provided in implementing the digital twin services.

2) The "data processing" criterion refers to the manner in which the data acquired in the physical world are processed in a specific digital twin application. Three sub-categories have been defined:

- statistical techniques;
- simulation methods;
- machine learning techniques.

Traditional statistical techniques refer to data processing methods based on analytical models. Simulation models include simulations of discrete events as well as finite element methods. Finally, machine learning techniques as well as other artificial intelligence techniques make it possible to implement more complex data processing. As indicated for data acquisition, the data processing classification criterion can also be represented on an axis with an incremental scale, which varies from a minimum performance level to a maximum level (i.e. from a statistical level to a machine learning level), depending on the complexity of its contribution to the digital twin's data processing capabilities.

3) The "safety problematic" criterion refers specifically to the safety domain of the digital twin application. It depends strictly on the main source of the risk generating the safety problems to be managed. This criterion has been divided into three main sub-categories that define the manner in which the origin of the risk is evaluated in the safety problem being analyzed:

- risks related to equipment;
- risks related to humans;
- risks related to human–equipment interactions.

The first category refers to the safety problems that chiefly involve risks due to the presence of hazardous equipment and processes. The second concerns safety issues that only involve operator tasks or interactions. The last category involves risks where operators and machines interact directly.

Figure 5.1 illustrates a 3D representation of the proposed framework. It is therefore possible to plot a scenario involving interactions between these three criteria for digital twin capabilities dedicated to industrial safety. Agnusdei et al. (2021) have proposed a state-of-the-art analysis that aims to apply this framework to the analysis of current practices. Their aim is to highlight the most critical characteristics of recent digital twin applications in the domain of safety.

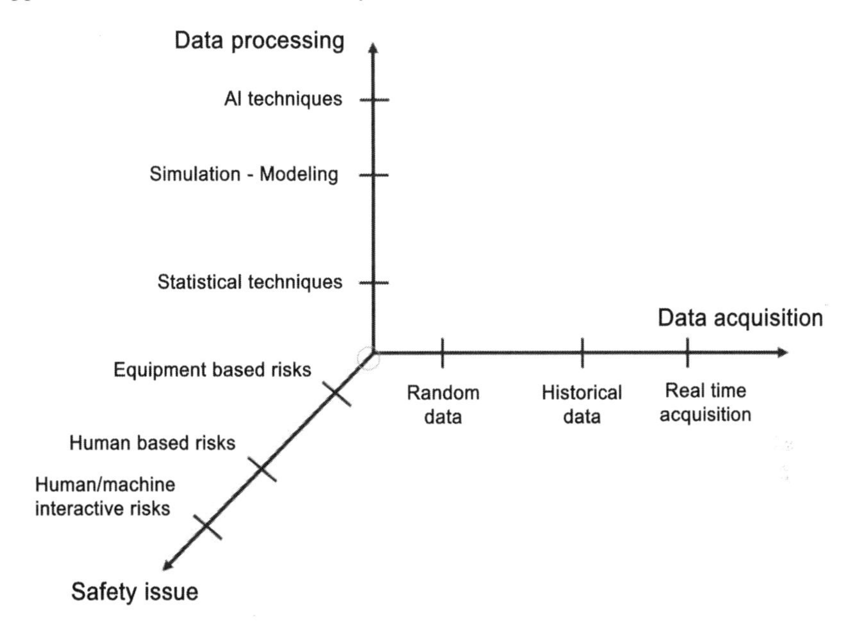


**Figure 5.1.** *The 3D representation of a framework to assess the capabilities of a digital twin used to enhance safety (Agnusdei et al. 2021)*

## 5.3. Modernization of a pilot installation of an ejector pump

In this example, the assigned goal is to experimentally simulate a situation that may occur during a petroleum extraction process. The pressure of a petroleum reservoir, whose value is higher than transport pressure, is used to create suction in the pool field, whose pressure is not high enough to allow transfer of oil into the transport network. In practice, the driving fluid is the liquid (oil) and the aspirated fluid is the gas. The pilot installation, an air and water cooled model, consists of a Venturi effect vacuum ejector pump that allows the transfer of a two-phase gas-liquid flow. Figure 5.2 presents the PID diagram of the ejector pump installation.

The pump, which is supplied by liquid from the open reservoir, produces a liquid flow under pressure at the input to the vacuum ejector. This component is considered the determinant in the installation. The depression created at the nozzle generates an aspiration of air. The two-phase, air–water emulsion is introduced into a vertical separator, where the liquid flows down to the bottom and the air rises to the head. The unit is equipped with two regulation loops, with proportional, integral and derivative control. The working of the divergent of the venturi extractor may be problematic in practice, because of the possibility of fouling or blocking with the petroleum fluids, resulting in overpressures. A pre-analysis of the potential risks led to the following list of events:

– explosion and blast in the ejector;

– flooding of the gas channel;

– flooding of the installation;

– explosion of the separator;

– inflammable vapor plume in the air.

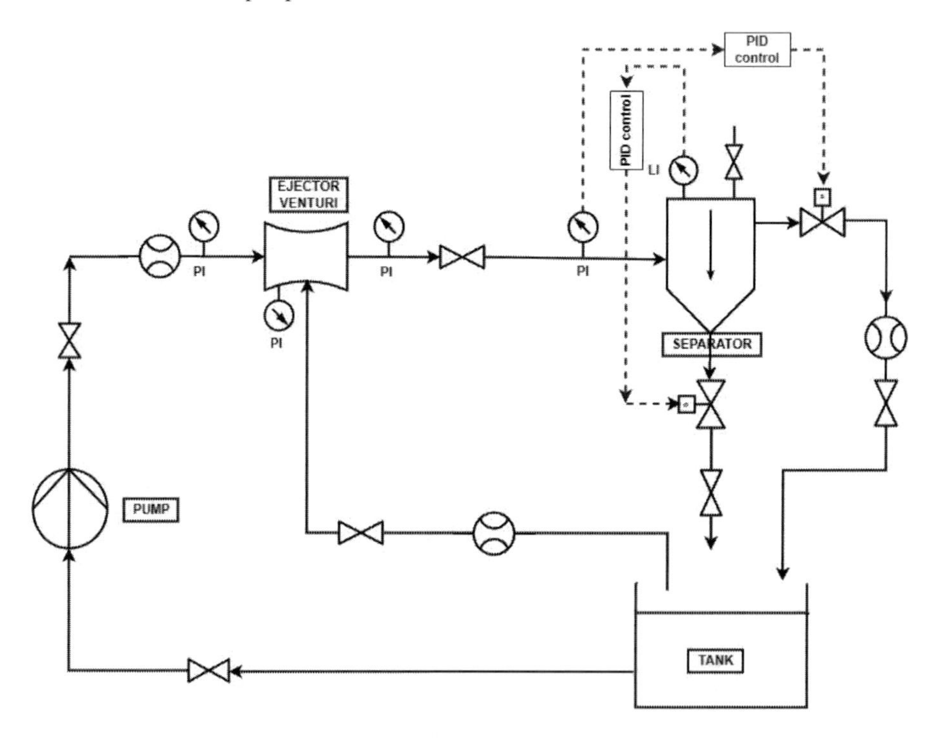


**Figure 5.2.** *PID diagram of the initial configuration of the vacuum ejector pump using the Venturi effect*

The goal of revamping the unit is to predict, prevent and protect against the risks identified, using smart solutions. The chosen solution was to create and use a digital twin. Figure 5.3 illustrates the PID diagram for the *reconditioned* configuration.

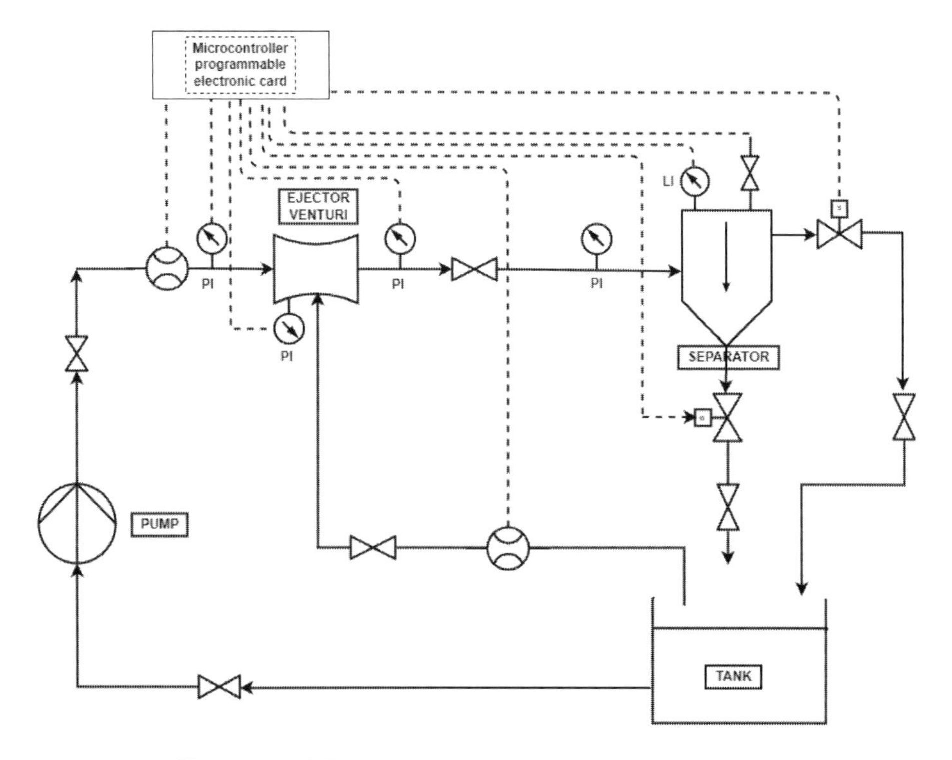


**Figure 5.3.** *PID drawing of the reconditioned configuration*

The new diagram makes it possible to observe that all sensors have been replaced by analog instruments. All data acquisition signals converge toward a micro-controller on a programmable electronic card.

The goals for the strategy to improve safety and maintenance conditions are detailed by Carlo et al. (2021). They are chiefly constituted by the respective creation of a digital twin model of the Venturi ejector, and a platform to rapidly detect the causes for malfunctions in the installation.

Table 5.1 summarizes the comparison of the criteria of safety, maintenance control, and access to the data between the initial and reconfigured installations in the sense of Industry 4.0.

## 5.4. Model for developing a digital twin to prevent OSH in the process industry

This example proposes developing a reference model for a digital twin dedicated specifically to the prevention of and protection against risks that the operators could be exposed to in process installations. The model was designed to make it possible to improve the level of safety of operators in their real working environment.

| Criterion | Initial installation | Reconfigured installation |
|---|---|---|
| Safety | Heightened risk for operators in proximity to the installation in case of malfunction | The simulation tool makes it possible to anticipate hazardous situations, by intervening in a preventive manner or by stopping the installation |
| Maintenance | Specialized operators must identify the zone in which to intervene in case of anomalies | Thanks to the anomaly detection tool, the time taken to identify the intervention zone is reduced considerably and no longer requires the presence of specialized operators |
| Control | The process can only be controlled in proximity to the plant and carrying out an overall assessment is very complicated | The platform enables real-time surveillance of the process using a web application; it is also possible to combine data from different sensors to obtain aggregated values that are representative of the situation |
| Access to data | Absence of data acquisition system | Data are acquired and stored in the cloud and easily accessible, locally and remotely. On the other hand, a system to protect against cyberattacks must be adopted |

**Table 5.1.** *Qualitative analysis of improvements resulting from the reconfiguration of the installation (Carlo et al. 2021)*

Bevilacqua et al. (2020) recently published the original reference model based on digital twin methodologies to contribute to risk reduction in process units, as shown in Figure 5.4.

This reference model must allow the company to:

– create a virtual process parallel to the physical process, which will offer a static and dynamic tool for analyzing the physical industrial process;

– propagate information to other, connected and realizable digital objects, in order to increase safety for the concerned actors;

– intercept anomalies from the start, in order to be able to quickly intervene to minimize damages following a break or failure, and to promote prevention and maintenance.

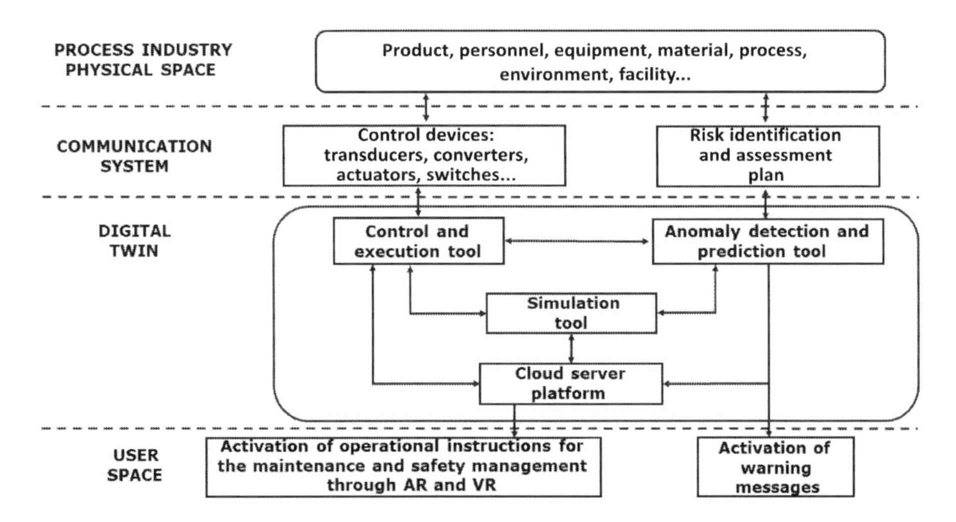


**Figure 5.4.** *Reference model for a digital twin used as a prevention support by process operators (Bevilacqua et al. 2020)*

### 5.4.1. *Description of the model*

The reference model is made up of four main layers:

1) The physical space of the industrial process: this layer includes all industrial physical resources such as the product, personnel, material, equipment, process, environment, installation, and so on. All observable elements that can be monitored, detected, controlled and activated are referenced in this physical space layer.

2) The communication system: the second layer is dedicated to all data and information that can be transferred between the digital twin and the elements of the industrial installation. All physical elements are monitored and detected using command devices and tools for executing data collection and to control devices with various appliances, such as sensors, cameras, actuators and other composite devices.

3) The digital twin: the third layer contains the conceptual digital twin infrastructure, which must carry out and manage the following tasks:

- acquisition of all data from the sensors in the installation;

- visualization of data from the installation;

- data analysis for anomaly detection and association predictions;

- integration of data from the anomaly detection algorithm to compare trends;

- development of the digital twin that simulates the behavior of the installation (possibility of developing simulation scenarios);

- integration of the digital twin and comparison with the physical space in order to determine abnormal behavior;

- identification of situations that are hazardous for the operator;

- activation and management of warnings on the 3D model transmitted toward the operators, unit managers and other system;

- maintenance support following anomalies.

The digital twin system is made up of four main tools: the control and execution tool, the simulation tool, the detection tool, and the anomaly prediction tool and the cloud server platform.

4) The user space: in the context of the last layer, the term "user" here denotes a person, a device or a system such as the manufacturing execution system (MES) and/or the enterprise resource planning (ERP) business software. Users are offered the interface through two solutions:

- the activation of operating instructions for maintenance management and system security management through augmented reality and virtual reality;

- the activation of warning messages for operators.

### 5.4.2. *Implementing the model*

The model is implemented in five interconnected sequences:

1) Phase 1: developing a plan for risk analysis and assessment: during this phase, particular attention is paid to assessing risks that could potentially impact operators as they carry out the process and during maintenance operations by integrating the influence of the presence of the digital twin.

2) Phase 2: developing a communication and control system: the specific identification of entities that establish a link between the physical reality and the digital twin is essential. Data collection and information gathering is supported through the use of existing connected sensors through the insertion of new sensors into existing infrastructure, based on automatons and controllers, through the addition of peripheral devices (IoT, wireless and wired sensor networks), and through the effects of actuators.

3) Phase 3: development of digital twinning tools: this phase focuses on the design and development of tools used in digital twinning. Machine learning algorithms have been used to detect and predict anomalies. The simulation tool is then developed at this stage, considering all the thermal, physical, mechanical, chemical, electrical, organization aspects, among others. The final step is that of designing the model for a real process and carrying out experiments with this model.

4) Phase 4: integrating tools in the context of digital twinning: the tools developed are then integrated into a dedicated software platform, which will carry out various activities of data acquisition, data manipulation, detection of state, diagnosing health, evaluating the prognosis and generating advice.

5) Phase 5: validating models and the platform: scenarios for process industries in which the reliability of the installations impacts operator safety and the efficiency of the organization has been used to validate the systems developed. This phase involves the verification of the functionalities of the platform and test activities. The test phase makes it possible to collect quantitative data in order to evaluate the performances of the chosen solution and to estimate the chief advantages, with particular attention paid to operator safety. The Return of EXperience is collected for revising/optimizing the tools and the platform.

### 5.4.3. *Conclusion*

Despite the quality and pertinence of this reference model for developing a digital twin that can be applied for ensuring operator safety, certain remarks must still be made. Thus, Bevilacqua et al. (2020) have highlighted two essential aspects related to the design and implementation phases of the digital twin, as well as certain practical implications of the proposed reference model.

As concerns the first point, the authors highlight the need to choose robust, resilient, stable, flexible and reliable solutions designed for real-time use, considering the reactivity of the operators. It is recommended that the targeted design of the human–machine interface be as non-intrusive or non-distracting as possible, and that it should have low energy consumption. Finally, various questions about normalization and standardization must be examined. In particular, these concern the synchronization of the physical object and its digital twin, the matching of syntax and semantic interoperability to ensure reciprocal collaboration of the digital twin and the safety of the physical object and its associated characteristics in terms of confidentiality and integrity.

The second point concerns the practical implications related to implementing cyberphysical systems. An important asset is related to the definition of a complex and integrated control system, which uses a network of cyberphysical elements, in order to ensure the continuity of the shared monitoring, even in the case of abnormal functioning of the installation itself, thus leading to a system of control and monitoring for and by operators.

## 5.5. Custom manufacture of food product by project development

This example proposes to study the plan to develop a process for the food industry, applicable to SMEs, proposing custom and on-request innovative technological solutions to produce granulated dog and cat food. Urban et al. (2020) first presented the schema for the principle underlying the plan to develop an automatic manufacturing and bagging line for kibble. Figure 5.5 depicts the generic example of the seven stages of the project.

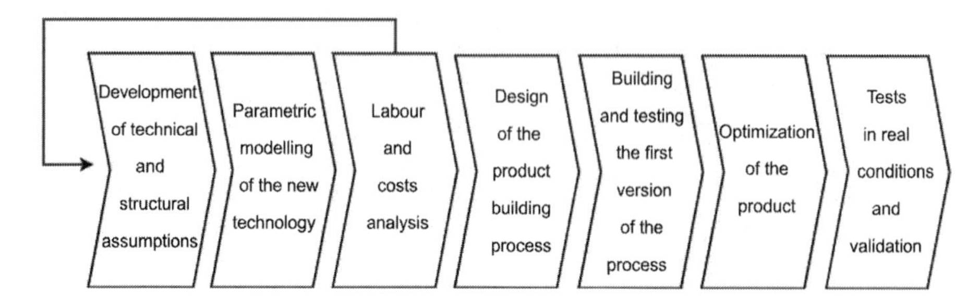


**Figure 5.5.** *Generic scheme of the principle underlying a plan to develop a process (Urban et al. 2020)*

The first step consists of developing the broad lines of the project as well as the technical and structural hypotheses. During this step, the specific tasks and functions that the machine must carry out are listed, the structure is defined and materials and components are selected. Based on these specifications, the developers created a parametric model of the device that uses new technology (step 2). The next step consists of analyzing the job tasks and the costs of the new technology being designed. This then leads either to the decision to go ahead with the implementation of the concept or to return to the hypotheses, as depicted by the backward loop.

Based on the modeling results, the company may start to design the manufacturing process for a new product. The models designed earlier only contained frameworks and specific elements of the technology, such that designing the implementation of the model would require enormous design work. Many sources (patent files, industrial shows and online services) have been explored to identify and verify possible solutions to the many technical problems detailed, including execution methods. Technical specifications have been developed to satisfy emerging technologies and the necessary materials and components are collected to move to the next step. Drawing inspiration from the digital Industry 4.0 technologies, Urban et al. (2020) first listed technologies that would help in improving each step in the project presented in Table 5.2.

The next step in the development plan consisted of constructing and testing the first version of the production line used to manufacture pet food. This task made it possible to discover structural errors that were not detected earlier and the necessary corrections were carried out. The tests took place in simulated working conditions. The penultimate step consisted of optimizing the developed product. Several modifications and adjustments were carried out at this stage. The final version of the product saw light of day and was installed within the client company and tested in actual working conditions. The technology was well validated in the context of the hypotheses that were initially established. The whole development project lasted 36 months.

| **Solution step** | **ID** | **S** | **VT/AR** | **VT/VR** | **PI** | **VM** |
|---|---|---|---|---|---|---|
| Developing structural and technical hypotheses | x | | | x | x | |
| Parametric modeling of the new technology | x | x | | x | x | x |
| Analyzing tasks and costs | | | | x | x | |
| Dimensioning the manufacturing process | x | x | x | | x | x |
| Construction and tests of the initial version of the process | x | | x | | x | x |
| Optimization of the product | x | | x | | x | x |
| Tests in real conditions and validation | | | x | | x | |
| Glossary | ID – Integrated database<br>S – Simulation<br>VT/AR – Augmented reality<br>VT/VR – Virtual reality<br>PI – Partner integration via digital tools and channels<br>VM – Virtual manufacturing | | | | | |

**Table 5.2.** *Digital solutions that could improve specific steps of the project (Urban et al. 2020)*

## 5.6. Impact of the design of a cyberphysical system on an industrial process

In this example, Jimenez (2017) transposed, for the process industry, the unified, five-level industry 4.0 architecture proposed by Lee et al. (2015) as the guiding principle for implementing a cyberphysical system used in the manufacturing industry.

A Dutch company own a factory that produces material in bulk using ores imported from overseas. The ore arrives at the port by ship. The transport from the port to the factory, on trucks, is managed by an external service provider. In the factory, the company mechanically transforms the mineral into powders and/or into grains of different types and quality. The factory has three separate installations:

– all primary material is stored in a large, empty hall;

– the production unit is a three-storey building, equipped with different machines;

– the vertical storage silo is where the final material, in powder or granular form, is stored.

Further, the operation of the unit is entirely automated, which means there is no standard work being carried out on the site, except for maintenance work and quality control. Finally, the installation aims to operate 24 hours a day, 7 days a week, for the whole year. The company's chief objective is to provide high-quality products to its clients. Indeed, this company offers personalized solutions through hundreds of finished goods, made as per the client's specifications.

### 5.6.1. *Choosing the problem to be studied*

After consulting with the stakeholders, it was decided that three main problems would be selected based on the interest level:

– an operational problem at the level of machines and equipment;

– a control problem at the factory level;

– a strategic problem at the organizational level.

The high noise levels is the major operational problem that companies have had to face in the past 2 years. The vibrations of the machine produce a low frequency that resonates throughout the building. This noise level prevents the company from investing in other machines and equipment. The factory comes under close scrutiny from environmental authorities, with the probability of warnings being issued, or even legal cases being filed and production being stopped. To try and resolve this noise problem, the company used paints and damping, noise absorption devices, which succeeded in partially reducing the noise. However, the problem still continues, as passive reducers are not capable of significantly reducing the sound levels attained.

The second major problem is the lack of reliable information for maintenance and quality control. At present, maintenance and quality control is carried out during the periodic examination but the company does not have any data management systems. Consequently, the team in charge of maintenance and quality control cannot learn from previous failures. Because of this, a storage assistance tool that can rapidly analyze failure modes will improve the factory's availability.

Finally, from a strategic point of view, there is a large disconnect between the suppliers, the production process and the clients. For example, at present, the client places an order by email, then the sales employees record the order in their internal system. Once the order is in the system, production reads the order, verifies that it is possible to fulfill it, and then passes it on to the production line. At the supplier's end, the company orders the primary materials from foreign suppliers through email. However, if the contents do not correspond to quality standards, the company must return the primary material to the supplier and use the stock of finished goods to fulfill the order. In conclusion, although the production process is highly automated, client relations and relations with suppliers are entirely carried out by and dependent on communication between humans.

### 5.6.2. *Design principle for the cyberphysical system*

A panel of 32 experts were asked for their opinion on the probabilistic maturity and impact of various Industry 4.0 technologies and this led to the technologies summarized in Table 5.3 being selected on priority. These allow *Key* and *Important* appreciations to be carried out for the project.

| Technology 4.0 | Maturity | Impact | Action |
|---|---|---|---|
| Artificial Intelligence | Average | Very high | Key |
| Sensors | Average | Very high | Key |
| Cloud storage | High | Very high | Key |
| Big Data | High | High | Important |
| Data routing and databases | High | Average | Important |
| Digital products | Average | High | Important |
| Portables | Average | High | Important |
| Websites and web applications | Very high | Average | Important |

**Table 5.3.** *Digital solutions chosen on priority to initiate the system design (Jimenez 2017)*

Jimenez (2017) set up a guide to design cyberphysical systems for the project. It is based on the implementation of the methodology proposed by Lee et al. (2015), which has a five-layer architecture. These layers are, respectively, dedicated to connectivity, conversion, virtual cyberspace, cognition and configuration.

"Connectivity" is the field that makes it possible to establish a link between physical objects, machines, robots and the data centers. This first layer has two aspects:

– a description of how the system will collect data in the physical world;

– the protocol for collecting and reading the data that are gathered.

"Conversion", at layer 2, signifies that the cyberphysical system can organize, convert and store information from the sensors in an intelligent way. This level is crucial, as the physical devices connected to the system send in data, which are difficult for analysis software and stakeholders to understand. Thus, there must be a converter to transform data coming from a source to information that is meant for decision-makers.

"Virtual cyberspace" aims to analyze, store and control the information supplied by the previous levels. This part of the system is essential, as it makes it possible for different stakeholders to access the information and to understand how the system behaves and communicates between the components and stakeholders. The four different components in layer 3 of the cloud computing system are the noise analysis software, the maintenance analysis software, the component for analyzing operational activities and communication storage.

"Cognition" occurs in layer 4 of the chosen architecture. Strictly speaking, cognition is the set of faculties and phenomena related to the functioning of the human mind. Here, cognition refers to a set-up that can learn and generate knowledge from past experience, or from the experience of other systems. It is thus possible to optimize the process and to prioritize solutions based on the objectives defined by humans. This capacity for auto-comprehension can only be produced using some artificial intelligence algorithms via learning processes, such as *machine learning* and *deep learning*.

"Configuration", which is the final layer in the architecture, focuses on translating the decisions and analyses from the previous layers into physical actions. This layer incorporates continuous feedback on information that the cyberspace provides to the physical system, with no human supervision or control. Indeed, the components in this layer act as a resilient system that regulates, replaces and prevents hazardous conditions across the system. To recall, the three major problems on which the system must immediately act are noise, maintenance and quality alerts and the active participation of stakeholders in the supply chain.

The complete details of the procedure described in the design guide for the cyberphysical system in the example discussed here are reported step-wise in the essay by Jimenez (2017). To demonstrate the pertinence of the approach applied to this example from the process industry, Figure 5.6 illustrates an overview of the impact of the new cyberphysical system on the functioning of the company, with respect to its strategic, control and operational aspects.

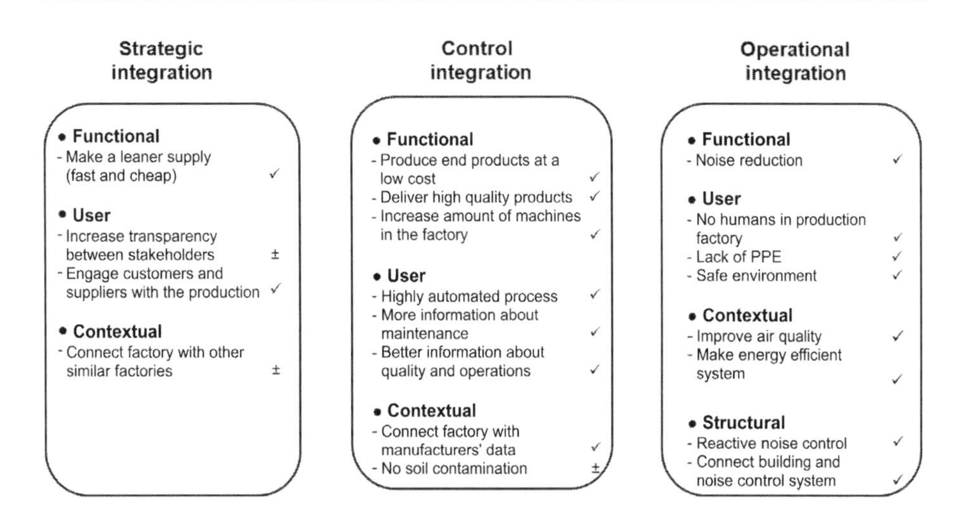


**Figure 5.6.** *Overview of the impact of the implementation of the cyberphysical system on the company's functioning (Jimenez 2017)*

## 5.7. Principle for redesigning a process in a cyberphysical production system

This example will show how it is possible to redesign a conventional process, with its existing equipment, through hybridization with a cyberphysical system (CPS), with the aim of obtaining a cyberphysical production system (CPPS) for a process revamped under Industry 4.0. Lins and Oliveira (2020) initiated a standardization procedure to try and promote this redesigning.

The authors first positioned the equipment systems based on the applicability of the industry 4.0 technologies. Figure 5.7 allows us to compare three industrial systems, traditional, automated and industry 4.0 systems, respectively, based on the degree of applicability of the different technologies concerned (zero–low–average–high).

Next, a reconfiguration process is brought in three parts: infrastructure, communication and application. The "infrastructure" part constitutes the basis for defining the specifications of the functional components and other leveling characteristics. The "communications" part must identify industrial networks present in the existing equipment and integrate them into the IoT network. In the absence of industrial networks, the communication integrates and manages only new IoT connections.

The "application" part consists of a generic modernization approach to supply and satisfy equipment with data, cloud computing and web services, as well as taking charge of initial applications that manage and control equipment. Table 5.4 summarizes the specifications of different parts of the redesigning process in terms of infrastructure (denoted by I), communication (denoted by C) and application (denoted by A).

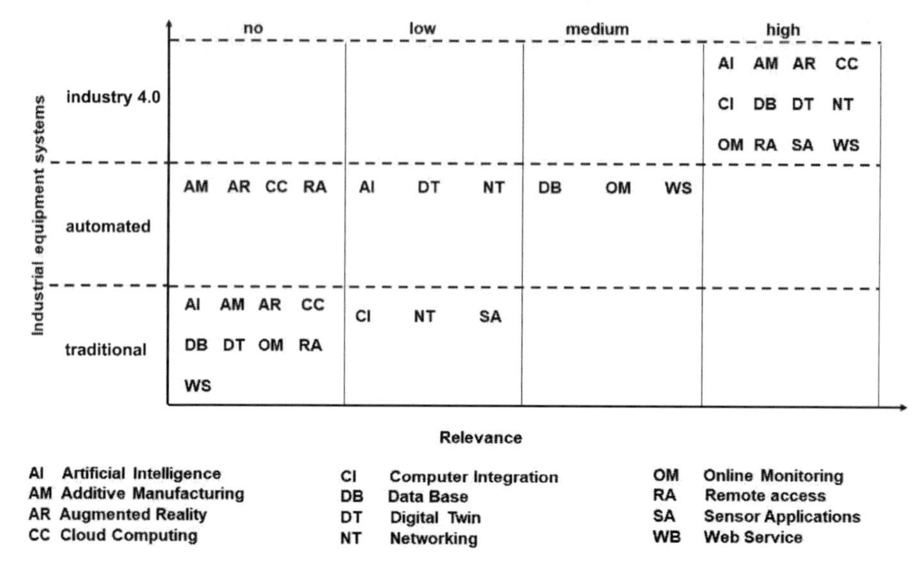


**Figure 5.7.** *Comparison of the different technologies in three industrial systems (Lins and Oliveira 2020)*

After having defined the specifications, the functional components of each part of the reconfiguration process must be detailed. These components are programs or devices with functions that support equipment to carry out Industry 4.0 functions.

In the "Infrastructure part" of the modernization process, the embedded systems are fundamental IoT components, which leads to their playing an essential role in Industry 4.0. They make it possible to carry out a low-cost reconfiguration, reducing the space used, enhancing connectivity and interoperability, reducing energy consumption and improving robustness. The two fundamental functional components are the EBoard card (Eboard-C) and IoT device component ((IoTDev-C).

In the "Communication part", in order for industrial equipment to communicate with Industry 4.0 technologies, communication components must carry out the integration of different communication technologies. The following functional components are considered:

– the network component (NET-C) is in charge of connecting industrial equipment with the Industry 4.0 network;

– the OPC component (OPC-C) includes the Server-M module, responsible for connecting the cyberphysical production system to the OPC servers and the Equipment-M module, to connect the CPPS with industrial equipment;

– the SDN component (SDN-C) is meant to respond to the demands of the SDN networks. This SDN-C component is made up of two modules. The first module is the Controller-M, which allows the CPPS to manage network resources, as well as to configure the data plan and exchange and transmit these. The second module, Commuter-M, allows the CPPS to receive data from an SDN control to decide on a pathway for the aggregated data.

| Number | Part | Specification |
|---|---|---|
| R01 | I | a survey of the needs and the improvements that can be brought to each industrial equipment or process |
| R02 | I | add IoT devices to industrial equipment, based on the functions of the equipment |
| R03 | I | add independent IoT devices that improve equipment, but are not directly installed on specific equipment |
| R04 | C | identify and map existing communication technologies and protocols of industrial equipment and Industry 4.0. |
| R05 | C | integrate existing technologies in the communication of equipment with existing networks in Industry 4.0 |
| R06 | C | integration of communication management, avoiding the use of a network manager for each type of communication |
| R07 | C | support for IoT networks, i.e., networks belonging to Industry 4.0 |
| R08 | C | studying real-time communication separately, as automated industry communication and the communication between Industry 4.0 components both take place in real-time |
| R09 | A | mapping of the software that already exists in the equipment, and mapping the requirements for its functioning in the production process |
| R10 | A | supporting the common applications of Industry 4.0, such as a data cloud, to collect equipment information and a web server for user access |
| R11 | A | integrating the existing software in industrial equipment into the new software used by the CPPS |
| R12 | A | using an application to monitor all information generated by the industry equipment in conjunction with the added IoT devices |
| R13 | A | support for remote access for users accessing the CPPS |

**Table 5.4.** *Summary of the specifications required for the cyberphysical production system (Lins and Oliveira 2020)*

As concerns the "Application" part, the functional components involving connectivity, accessibility and availability are:

– the component (DB-C) connecting the database to the CPPS;

– the remote access component (RA-C);

– the web service component (WEB-C);

– the monitoring component (MON-C);

– the cloud computing component (CLOUD-C) with the Node-M and Migration-M modules.

After the specifications, components, functions and technologies are defined in this way, a reference, i.e. Industry 4.0, architectural structure must be used. The RAMI 4.0 architecture, presented in the 2015 edition of the Hanover Trade Fair (D) was chosen, given the pertinence of the three-dimensional model of the elements involved, bringing together organization, use and interactivity (Hankel and Rexroth 2015).

| Number | Method | Component | Specification |
|---|---|---|---|
| 1 | CPPS retrofitting pre-configuration, defining the resources for industrial equipment | All | R01, R04, R09 |
| 2 | Infrastructure installation with embedded boards and IoT devices | EBoard-C IoTDev-C | R02, R03 |
| 3 | Identification of IoT devices and their packages | EBoard-C IoTDev-C | R02, R03 |
| 4 | Identification of OPC servers and industrial devices | OPC-C | R05, R06 |
| 5 | Identification of the SDN network, with the switches and network resources | SDN-C | R05, R06, R07 |
| 6 | Integration of IoT devices and industrial devices | EBoard-C IoTDev-C | R05, R06, R07, R08 |
| 7 | Detection and establishing of connection with Industry 4.0 communications | NET-C | R05, R06, R07, R08 |
| 8 | Online monitoring of all devices connected to the CPPS | DB-C WEB-C MON-C | R10, R11, R12 |
| 9 | Sharing CPPS information | DB-C WEB-C | R10, R12 |
| 10 | Installation/migration of native industry applications in the cloud | CLOUD-C | R10, R11 |
| 11 | CPPS remote management and maintenance | RA-C | R13 |

**Table 5.5.** *Methods used in the CPPS retrofitting platform (Lins and Oliveira 2020)*

Lins and Oliveira (2020) give a detailed description of the organigram for the reconfiguration process, how it can be integrated into RAMI 4.0 architecture and the functional reconfiguration platform. As an example, Table 5.5 summarizes all the methods used during the implementation of the procedure on the platform to respond to the need for specifications and the functional components of the reconfiguration method.

This methodology was experimentally validated through the reconfiguration of an old automatic arm into an industrial prototype robotic arm with six degrees of freedom. Several essays comparing the initial equipment and its reconfiguration have demonstrated the efficiency gained in response times in real-time communications, as well as energy savings.

## 5.8. Systematic integrated approach to improve the processing of contaminated sediments

The contaminated sediments cause health problems in ecosystems and populations living close by. Heavy metals, polychlorobiphenyls (PCB) and polycylic aromatic hydrocarbons (PAH) are the main components. The process to manage the contaminated sediments requires the implementation of different, complex biological, physical, chemical and thermal treatment processes. In the majority of cases, the contaminated sediments are first mechanically, hydraulically or pneumatically dredged, then transported, stored and treated ex situ. The goal of the treatment processes of contaminated sediments is to extract, immobilize, destroy or neutralize the contaminants.

### 5.8.1. *The Novosol® process*

The Novosol®, process, developed by the Solvay SA group, combines different treatment techniques in order to stabilize mineral residues polluted by heavy metals and organic components. The process is divided into two steps: a phosphation step followed by a calcination step.

The phosphation phase makes it possible to ensure the stabilization and reduction in solubility of the heavy metals in the sediments by adding phosphoric acid. The calcination phase makes it possible to reinforce the stability of the metals resulting from the phosphation step and thermally destroy the organic components. The operational details of the process are described in the work published by Hardy (2010).

The chief advantage of the process resides in the valorization of the treated residues through their incorporation into road-building materials, destined to build pavements.

### 5.8.2. *The sociotechnical Novosol® system*

The system is made up of different operators and actors who initially intervened during the development phase of the process and who now intervene in the operating phase. At the client's request, the sociotechnical system was limited to the subsystem of the phosphation step.

Hardy (2010) and Hardy and Guarnieri (2013) proposed applying the dynamic integrated STAMP model, centered on system security, to the Novosol® process. This approach is not limited to the strictly technical point of view. It promises a global overview of the process, considering all interactions, including human and organizational factors.

The analysis begins with a list of all the defined variables that make it possible to identify all interactions in the global system. A static safety control model is then established to visualize the organization of the system as well as the interactions within the system. The steps then continue with the dynamic model looking at the system behavior, especially during the transition from normal operation or a secure state toward a degraded state. This dynamic phase involves the dynamic modeling of the system and the formulation of the results and recommendations as happens during a REX process.

The final objective of the routing of the analysis procedure consists of assessing the system security by progressively deploying the systems-theoretic process analysis (STPA) technique or integrated hazard analysis, both for the static as well as dynamic process.

### 5.8.3. *Conclusion*

This example highlights how applying the STAMP model and the integrated hazard analysis STPA technique improves the safety of a sociotechnical system. The methodological illustration of its use in the case study for the industrial Novosol® process for remedying the treatment of contaminated sediments has shown the pertinence of this global approach by not limiting ourselves to only the technical aspect as is conventionally done.

## 5.9. Digitalization to benefit safety management

Putting in place a safety management system (SMS) founded on a systemic approach makes it possible to implement a managerial approach to safety, bringing together all the personnel in the enterprise by seeing their participation and incentivizing their involvement. The SMS constitutes an organized and coherent whole, allowing the assessment, control, correction and improvement of all elements belonging to the field of industrial security, especially the field of process safety. However, it can also be applied, through specific extensions, to the prevention of occupational hazards (health), to industrial hygiene and to environmental protection. The content of an SMS must define, as coherently as possible, a company's policy and for each factory or establishment, it must define the objective values and targets, the organization, plans of action and means of implementing them, the actors and their behavior and training, as well as the display, diffusion and valorization of the obtained results.

Jones (2019) and Jones and Menon (2021) proposed a safety management approach integrating the advantages brought in by digitalization techniques. They sought to promote a new, emerging category of operational hazard management software using the following:

– offering desktop web applications and mobile applications that can be used on the ground that make it possible to carry out risk assessment, barrier management, safety management, permitted activity authorizations, management of change (MoC), incident reports, etc.;

– acquiring, processing, storing and archiving data coming from various devices on the ground via IoT;

– generalizing the creation of active dynamic digital twin models to enhance advanced modeling capabilities;

– promoting potential big-data analysis to provide usable information on states and trends in system hazards.

The unique feature of this new approach for a cumulative estimation of all risks in the installation consists of modeling their impact on the groups of barriers in the processes and associating them with the management of risk of accidents. Figure 5.8 depicts the core of the proposed model. The software is fed data from a model that represents clusters of barriers subject to deviations, discrepancies and non-conformities identified in the installation (Figure 5.9). For example, the classic deviations with respect to standard functioning generates functional big-data extracted from the system, which can be categorized into:

– delay in maintenance;

– delay in inspections;

– non-occurrence of planned maintenance;

– high-risk working conditions;

– inspection failure;

– feedback from observation on the ground;

– management of change;

– incident reports;

– state of critical equipment;

– audit results;

– gaps in competencies of teams;

– deviations from operational limits (integrity operating windows [IOW]).

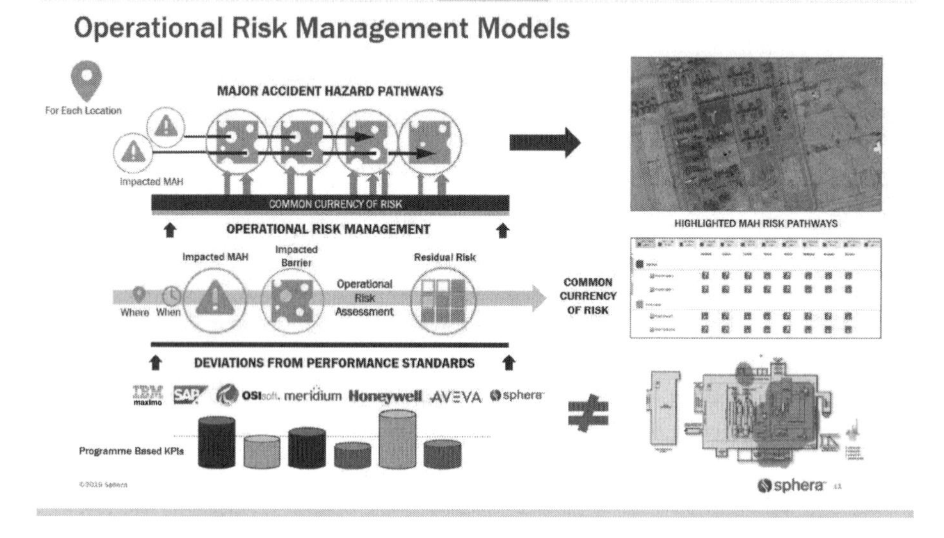


**Figure 5.8.** *Core of the risk management model (Jones 2019)*

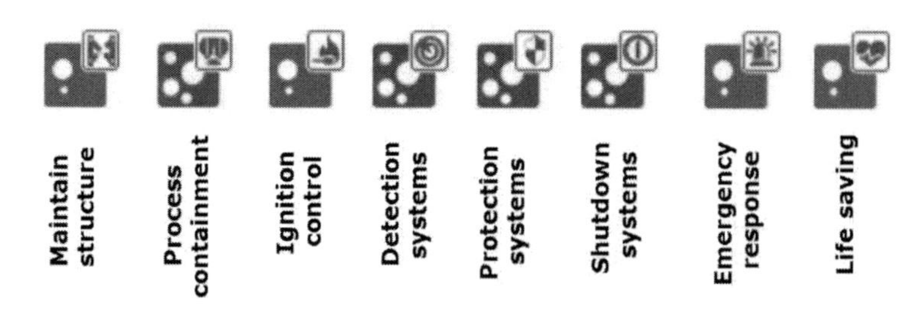


**Figure 5.9.** *Cluster representation of barriers (Jones and Menon 2021)*

Jones (2019) reported two examples of the application of this approach using an operational risk management platform, implemented by two international oil and gas industries.

### 5.9.1. *Improvement in the quality of technical risk assessment and modeling the impact of cumulative risks*

A large, international oil company has several off-shore platforms on the United Kingdom continental shelf. This operator has a mature management system and a clearly defined approach to process safety, to manage the integrity of the assets and to monitor work. By implementing the solution using the operational risk management platform, the technical division sought to improve the risk assessment for situations where critical equipment no longer meets performance standards. The existing practice consisted of carrying out an operational risk assessment as soon as such a deviation was identified during an inspection, maintenance or a formal review of the operator's activities. This initial risk evaluation was approved by the local manager in the offshore installation and was discussed with the onshore technical support team. A general criticism of the external regulatory body in the sector (which was not specific to this company) was that in practice these kinds of risk assessments rarely identified the real hazard associated with the failure in the safety function in the critical equipment. This company had a clearly defined approach to managing elements that were critical for safety as well as for the associated components and equipment. A performance standard was defined for all the components identified in each installation based on the risk reduction constraint imposed in the regulatory safety dossier. In order to improve the quality of risk assessment, the site manager used operational management software to present the evaluator with estimations from risk assessment models based on the type/category of each modified element. These models made it possible to:

– define the real danger that could be posed by non-conforming equipment as a category;

– provide the typical attenuation measures to take into consideration in order to minimize the risk based on the class of equipment/function;

– present content relevant to the performance standard for each element in the form of control lists, which encouraged the evaluator to:

- identify the criticality level for safety or integrity associated with the deviation presented,

- consider other protection functions that could aggravate the problem, that is, other deviations that also have an impact on the zone and the major hazard under control.

The new software also helped the evaluator in deciding whether the deviation would have an impact only on a local zone in the installation, or on the whole of the platform. For example, a single gas detector may have an impact on a localized risk, while fire water pumps that cannot provide the required capacity of water could affect the whole of the platform.

The software was also used to steer a revised approval process:

– formally involving the technical authority defined in the approval for the technical risk assessment based on the equipment class/function;

– indicating the approvals required from operators of assets and activities, in addition to the local operator of the installation based on the criticality level in terms of safety or integrity as identified in the assessment.

The operational risk management software can also be used to manage all the activities the operator is authorized to carry out on the installations. To summarize, it is thus possible to obtain a combined view of all risks related to the equipment and all risks associated with the activities on a barrier model by highlighting all paths through which the hazards could propagate.

### 5.9.2. *Providing a real-time view of the actual state of critical equipment and their impact on the risks*

A large oil company will construct and operate a world-class refinery in the Middle East. Presently under construction, the company has chosen an Industry 4.0 initiative to develop and provide a technological approach that aims to integrate a series of enterprise systems to improve the development of processes and the management of assets and operations.

The company has a sophisticated safety management system and clear corporate standards and practices. It wishes to have a real-time view of the risks associated with operational safety and processes and for this to be the core element in decision-making as soon as the factory becomes operational. The operator is putting in place the operational risk management software platform to manage all activities and deviations authorized on the installation. This software will be integrated with three other systems within the company to provide a real-time view of the state of risks associated with critical equipment:

– a (*data historian*) server, with a near-real-time status for critical equipment, critical alarm systems and the digital control system;

– a maintenance management system to log inspection and maintenance, and the associated plans and schedules;

– an operator patrol system.

As the project is still in the design phase, the project team could access the design documents from external service providers in charge of each unit in the refinery, to identify the critical equipment that will be in these units. This chiefly involves bringing together a variety of information in useful formats:

– from the design phase, the HAZOP studies and studies on the integrity of the assets;

– the health parameters associated with specific components in the identified critical equipment;

– the types/categories of critical equipment associated with the process safety barriers model;

– logs representing all the critical equipment entered into the software.

Thanks to this integration, the software constantly monitors the state of the critical equipment, based on the above information and parameters, through the following sources:

– the state of the equipment in near-real-time, obtained from the data historian;

– inspection logs for critical equipment, from the patrols carried out by operators and the inspection management system;

– the deferred planned maintenance for critical systems, from the maintenance management system;

– associating the non-conformities with the process safety barrier model.

As in the previous example, the software is used to manage all authorized activities on the installation and the integration described here offers a composite view, in real-time, of all the risks.

## 5.10. Detection of deviations in the functioning of a heat exchanger through an artificial neural network

An artificial neural network is a particularly interesting tool for studying process applications real time. Himmelblau (2000) illustrated how a direct-action neural network could be used to detect operational faults in a heat exchanger. This example looks at the case of detecting deviations due to clogging and fouling in a thermal exchanger, using an 8-12-3 neural network (147 parameters) trained with $4 \times 80$ data. The heat exchanger is a shell-tube heat exchanger, with countercurrent fluid circulation. The bundle of tubes is made up of 108 tubes, arranged in a 4 cm square pitch, 4.88 m in length, with a diameter of 3.18 cm, which are initially clean. Table 5.6 lists the various operational data from the exchanger.

| Data | Cold fluid | Hot fluid |
|---|---|---|
| Fluid | Mixture of water (55%), ethylbenzene (25%) and styrene (20%) | Water |
| Flow rate (kg/h) | 9,080 | 4,086 |
| Inlet temperature (K) | 533 | 977 |
| Outlet temperature (K) | 708 | 599 |
| Inlet pressure (kPa) | 207 | 345 |
| Outlet pressure (kPa) | 148 | 236 |

**Table 5.6.** *Pre-specified data for normal operation of a heat exchanger (Himmelblau 2000)*

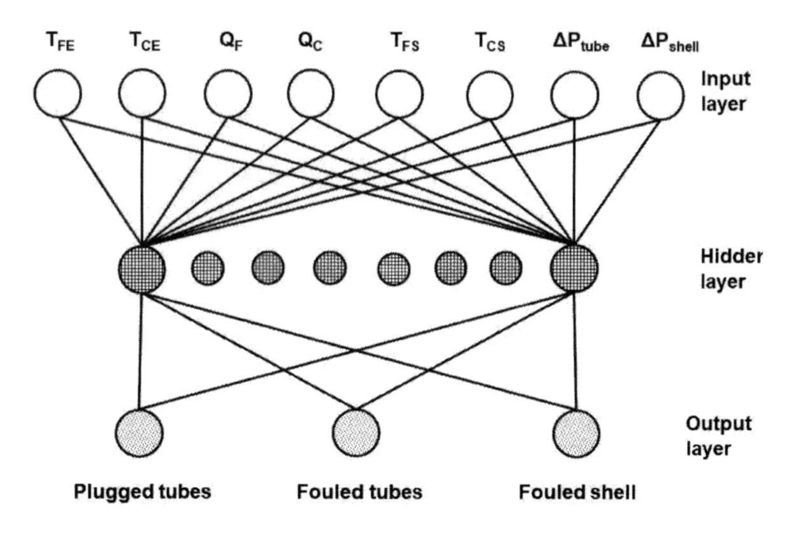


**Figure 5.10.** *Architecture of the neural network used to train for the detection and diagnostic of faults and deviations in the operation of the heat exchanger (Himmelblau 2000)*

A simulation code was used to generate data that are representative of the cleanliness and fouling in the heat exchanger, operating in a continuous, countercurrent regime. A specific file with data for each normal condition and each defective condition was prepared. Information on the thermodynamic properties of the fluids, the physical properties of the fluids and the exchanger, the configuration of the heat exchange, the number of tubes, the size of the tubes and the shell, and the thickness of the fouling on the walls of the tubes and the shell are pre-specified. The deviation in terms of clogging in the tubes is described by four categories indicating the respective numbers of the clogged tubes, that is, 1-2-3 and 5. The deviation in terms of fouling in the tubes is described by four classes of the reduction in the straight section available to the fluid, that is, 1-3-5 and 8%. Two conditions of noisy data, distributed normally with a significance level of 0.01 and 0.02, are associated with the flow rates, temperature and pressure loss in each of the two fluids. A neural network training method used each input dataset from the normal mode as well as in degraded mode, pre-formatted with normed values between −1 and +1. The training is such that the value 0.9 represents a normal state, while the value 0.1 indicates a degraded operational state of the heat exchanger. For example, a target output configuration of 0.9-0.9-0.9 represents a case where the heat exchanger is clean. The result 0.1-0.9-0.9 illustrates a case where one or more tubes is clogged and so on. To test the neural network, a threshold value of 0.5 for the value of an output node was used as a discriminating criterion for the classification. If the activation of the output node was greater than 0.5, then this node was considered as being activated and represented a defective state in the heat exchanger. On the contrary, if the value of the node was less than 0.5, then the node was considered non-activated and no fault was estimated to have occurred. An example of the results obtained in the case of fouling in the tubes is given in Table 5.7.

| % reduced area | % correctly classified | |
|---|---|---|
| | Training | Test |
| 1 | 97.50 | 86.25 |
| 2 | 98.75 | 92.50 |
| 3 | 100 | 93.75 |
| 5 | 100 | 95.00 |

**Table 5.7.** *Example of the neural network classification, for fouling in the tubes (when the noise variation coefficient was 0.02) (source: Himmelblau (2000))*

Section 6.5.3 in Chapter 6 will emphasize that quantitative risk assessment tools based on the availability of digital data must be under constant vigilance. Thus, in this example, making use of the neural network must be systematically questioned. Other methods must be tested before using this, for example, the classical multivariable regression method. In the heat exchanger example, Himmelblau (2000) has also simultaneously used two other methods: Bayesian classification and the nearest-neighbor method. The author cautiously concluded that testing multivariate hypotheses on averages of the measurements gave much larger classification error rates. From this, they deduced, curiously, that results from neural network analysis were not worse than those obtained using traditional classification methods.

## 5.11. RFID applied to the prevention of occupational hazards

Radio Frequency IDentification (RFID) is a generic term that denotes a vast group of applications used to identify objects, in the broader sense, using radiowave communication (wireless technology) (Tihay 2012). Any object bearing a tag (people, animals, objects, plants, etc.) can a priori be identified by RFID wirelessly and without contact.

RFID is based on the principle of information exchange between an interrogation device and a transponder. The electromagnetic field generated by the interrogation device serves as the vector of communication between this device and the transponder.

### 5.11.1. *Fields of application of RFID technology*

There are two main fields of applications with, respectively, the tagging of objects and tagging with direct or potential reference to living things (Moustaine and Laurent 2012; Tihay 2012).

#### 5.11.1.1. *Applications related to tagging objects*

The fields of applications for tagging objects are related to:

– logistics and product traceability, with identification and localization of reusable goods and merchandise, etc.;

– industrial production of goods, the monitoring and maintenance of processes, with archival systems, equipment management, monitoring of the environment, etc.;

– safety, quality and tracking of products.

#### 5.11.1.2. *Fields of application of tagging with direct or potential reference to living things*

Fields of application for tagging living things are related to:

– access control and traceability of people and animals;

– loyalty cards, membership cards, payment cards;

– the health sector, with the management of medical material, its cleaning and recycling, the traceability of bags of blood, etc.;

– the field of sports, hobbies and domestic goods;

– public services, with remote reading of electricity, gas and water meters, the management of smart garbage bins, integration of the technology into identity documents, etc.

### 5.11.2. *RFID applied to occupational safety and health*

RFID technology can improve occupational safety and health (Tihay 2012).

#### 5.11.2.1. *Application to machine risks*

RFID can be used in the context of reducing hazards related to the presence of machines. It is possible to distinguish two categories of systems based on whether or not they display the status of the safety component. In the absence of the status of the safety component, a warning message is issued, but with no guarantee of its efficacy and its availability is not guaranteed. In the case of a safety component, the manufacturer guarantees its fitness for the function and its safety level, for example, in the process industry, using a Safety Integrity Level (SIL).

#### 5.11.2.2. *PPE management*

The chief objective of PPE is to enhance operator safety in the industrial environment. Personal protection equipment, such as jackets, clothing, helmets, noise-cancelling earphones, glasses, masks, safety shoes, etc., must be worn appropriately in accordance with specific regulations. These equipment can be identified using RFID chips integrated into the PPE. These sensors are thus able to signal the presence of an operator equipped with PPE, and can also send other useful data such as temperature, humidity levels, orientation, duration of use, etc. (Sole et al. 2013).

#### 5.11.2.3. *Access control and securing hazardous zones*

An RFID system can contribute to risk reduction when it is used as an access control device. Entry into a hazardous zone or working on a machine may be restricted by putting in place a lock that contains an RFID interrogator. This can only be unlocked downstream of a transponder, which will authorize access after first verifying the safety requirements. The transponder often consists of a badge, which can also be integrated into PPE (Musu et al. 2014; Cabreira et al. 2015; Chatouane 2015).

#### 5.11.2.4. *Detection and location of people*

The principle behind detecting people is based on RFID and is used to prevent collisions between mobile vehicles and pedestrians. The RFID interrogator installed on the rolling device generates an electromagnetic field that defines a detection zone around the vehicle. The efficiency of detection does, of course, depend on anyone entering that zone wearing a transponder. RFID also makes it possible to locate the presence within one or more given work zones. Several interrogators are positioned such that they define distinct zones in the workplace. Each operator wearing a transponder can thus be located, in real-time, in whichever zone they are present. By extension, it is also possible to analyze different work posts within the workshop and use the accumulated data to deduce zones that present risks (Helmus 2007).

## 5.12. How RFID contributes to industrial engineering safety

Industrial engineering consists of all activities that make it possible, for example, to transform a chemical manufacturing process into an industrial production site. Engineering activity includes, among other things, several steps, including design, supply and construction, before the unit is effectively operational. This example aims to show how RFID can contribute to a potential improvement in safety during the design and construction of an industrial unit (Behesti et al. 2015).

Accident data from the chemical process industry indicates that 73% of accidents are likely to be caused by technical and engineering failures (Kidam et al. 2010). A list of characteristic features of these accidents revealed:

– Failures in auxiliary systems or their components, such as the piping systems, are a typical example of these. Their integrity and reliability depends on many factors, including design, complexity and management. It is therefore essential to choose simpler designs and more robust materials for construction.

– Typical design flaws in plant design, such as the use of inappropriate material for equipment, incorrect design, inexact specifications, poor implantation and defective arrangements.

– Operational failures resulting from the non-respect of optimal limits for original technical specifications or poor change management, with a lack of archiving and updating of technical dossiers.

The RFID system can be used to create links between construction and design, using the system's ability to develop a dynamic database for construction activities. It is thus possible to ultimately reduce the time required for the construction step, improve the safety on the site and qualitatively improve the timeframe for construction. Table 5.8 thus illustrates the example of how RFID can improve engineering safety.

It is useful to mention here that safety, which is an important factor in the construction of an industrial unit, is enhanced through the use of RFID. It makes it possible to reduce construction time, and, consequently, to reduce the associated risks. Furthermore, it creates an adequate tool for the project management tool, so that the team can plan for the smallest possible number of interactions between the construction team and the activities to reduce the sources of danger.

## 5.13. Exploring the idea of a socially safe and sustainable workplace for an Operator 4.0

Work and research on Industry 4.0 have chiefly focused on machines and did not consider, or adequately consider, the role of human beings in the design and operation of a smart factory. However, factories are not only made up of machines, but also of human beings, operators and workers, who cooperate with the machines and collaborate with each other in various ways, for example, executing tasks, monitoring the process, loading or unloading machines, interacting with machine interfaces, etc. Despite the increasing automation of production lines, humans continue to play a central role in the monitoring of production process and the execution of delicate, complex and strategic tasks. Consequently, humans remain in charge of the increased productivity of factories and high quality of the products.

| Activity | Traditional method | Using RFID |
|---|---|---|
| Equipment transfer to construction site | Equipment depot in the wrong position<br>Possible interference with other activities on the site<br>Transferring the wrong equipment<br>Waste of time in finding the right equipment | Right depot position<br>No interference<br>Project schedule checked<br>Right TAG – correct equipment<br>Finding the equipment in the right location |
| Equipment installation and progress report for piping activities | Delay in equipment report<br>Delay in preparing piping spools<br>Possible weld repair requirement<br>Scaffolding not available | Automatic report<br>Showing the list of connecting pipes<br>Proper fit-up and welding activities<br>Scaffolding schedule respected |
| Piping installation activities | Delay in piping spool activity<br>Waste of time in finding the pipe spool in the piping yard<br>Delay in piping sandblast and paint activity<br>Scaffolding not available | Appropriate schedule for piping spool activities<br>Find the piping spool in the right position in the yard<br>Right schedule for piping and sandblast activity<br>Right schedule for scaffolding |

**Table 5.8.** *Contribution of the RFID method compared to the traditional method (Behesti et al. 2015)*

Peruzzini et al. (2020) proposed combining physical data and virtual objects, digitized through the IoT (i.e. using digital twins), not only for machines but also for people. The factory's digital twin is constructed using data from the sensors and makes it possible to contextualize information to create self-adapting systems that are able to intelligently adjust production schemes based on different fields of application, also including human data in the creation of the factory's digital twin. The authors thus defined both a theoretical framework to introduce human factors into Industry 4.0, as well as a structure procedure to carry out a pragmatic assessment of the relation between measurable physical and cognitive human factors, and the design of the worksite. Indeed, sustainable industrial systems must no longer satisfy only performance-related goals, such as cost, quality, speed, productivity, flexibility, or adaptability, but also objectives related to humans, for instance, ergonomics, mental strain, intuitiveness of actions, ease of use of tools, satisfaction and efficiency. Consequently, the new characteristics of connectivity and the interoperability of systems and machines must be coupled with these sustainable development goals in order to assimilate data related to the human being at the level of the smart factory, and, at the social level, to promote the design of more sustainable, smart and flexible manufacturing processes.

In practice, the approach to defining the "Operator 4.0" framework is based on three steps:

– creating the digital twin of the real factory;

– monitoring the human physiological responses of the workers;

– assessing human factors.

An Operator 4.0 model must be constructed, expressing the understanding of the human–system interaction by monitoring the physical and cognitive workload of the workers using objective and subjective measurements. In order to do this, the installation has a biosensor to track physiological parameters, an oculometer to analyze visual interactions and a motion capture system to analyze physical movements. This is complemented by subjective questionnaires. A stress analysis is not included in this study due to the complexity of integrating measurements related to stress in an industrial context. The hybrid human–machine model was tested on an assembly work post in the industrial and agricultural vehicle sector in collaboration with CNH Industrial. A review of the application of the Operator 4.0 and the Return of EXperience revealed the following observations:

1) the 3D immersive virtual configuration allowed the concerned operators to assess the virtual work post, created to the scale of the real industrial site, to immerse themselves in the virtual scene and to carry out the same functions as the other operators and to adapt to the situation;

2) the human tracking system contributed information that was useful for evaluating posture and made it possible to detect particularly critical conditions, which had never been analyzed earlier;

3) the adoption of human monitoring tools, at the level of the workshop, was not judged to be too intrusive and the operators were positive about it as they felt safer and better monitored;

4) tracking of movements through optical cameras, however, was not always possible in the workshop because of light interference and calibration problems. It was only used in experimental tests in the laboratory;

5) data about human interaction could be managed in a valid manner through a IoT architecture so as to include the surveillance of vital parameters within the framework of Industry 4.0.

In conclusion, the application of the hybrid man-machine digital twin demonstrated how the proposed approach could efficiently support the simulation of the human–machine interaction so as to identify critical work conditions, improve the workers' perception of comfort and avoid ergonomic problems in the workshop.

## 5.14. Industry 4.0 challenges related to safety and the environment in the leather industry

This example aims to demonstrate how the application of digital Industry 4.0 technologies could be adopted for the leather industry in Bangladesh, in order to try and make production and working conditions healthier. The tanneries sector is made up of around 220 installations, half of which are family-run businesses (Deur 2020). The tanneries represent the most heavily polluting industrial sector, polluting water, air and soil, and are also most harmful to health and occupational safety. Multiple chemical operations are involved from the collection of skins and raw hides of domestic animals to the finished leather. These lead to the production of vast quantities of environmental pollutants, such as solid waste (skin, hair, lime, chrome residues, clippings, etc.), liquid waste from the leather, containing heavy metals, and gases like hydrogen sulfide, ammonia, chlorine, etc. (Dixit et al. 2015). Consequently, concepts from Industry 4.0 and Safety 4.0 can be used to control and minimize this environmental pollution and help business in the leather industry improve their operational practices by adopting automation and cleaner technologies.

The main challenges in implementing Industry 4.0 digital technologies are reported in literature. Table 5.9 lists 10 challenges, with a short description of each, that impede or complicate the practical application of these technologies.

Moktadir et al. (2018) used the BWM (Best Worst Method) to carry out a multicriteria decisional analysis in order to establish the mean optimal values of comparison of the respective challenges. A panel of eight groups of experts, identified by the numbers 1 to 8 and consisting chiefly of production directors, operation managers, supply chain managers, logistics managers, and technologists from the leather industry, categorized the different challenges based on their importance, as per the expert opinions. Table 5.10 summarizes their choices by choosing a series of difficult challenges and a series of less difficult challenges, respectively.

The detailed procedures for the selection of criteria and the process used to analyze them through the BWM are described by Moktadir et al. (2018). Table 5.11, which illustrates the final result, makes it possible to examine the optimal average weight for each challenge, and to carry out a comparison using this. Challenge 3, "lack of technological infrastructure", is the most crucial of these challenges as it concerns implementation of Industry 4.0 digital techniques in the leather industry. Challenge 8, "the complexity of re-configuring the production pattern", is in second position, indicating the weight of this challenge inherent to incorporating recent, automated technologies into smart manufacturing practices. These two large challenges are followed by "data uncertainty" (number 1 in the table) and that of high investments (number 2 in the table).

| Number | Challenge | Brief description of the challenge |
|---|---|---|
| Chal. 1 | Data uncertainty | Lack of systems to ensure enough data protection for the manufacturing companies |
| Chal. 2 | High investment | Industry 4.0 initiatives in the manufacturing industry require huge capital investment |
| Chal. 3 | Lack of technological infrastructure | Non-existence of technological infrastructure to support the manufacturing company implementation of Industry 4.0 |
| Chal. 4 | Unstable connectivity among companies | Non-secure connectivity impairs real-time communication among manufacturing companies, complicating the implementation of Industry 4.0 |
| Chal. 5 | Decrease in job opportunities | Industry 4.0 implementation in manufacturing industries takes away some jobs due to the replacement of humans with robots and the intense use of automation in the production system |
| Chal. 6 | Lack of strategy for Industry 4.0 | Lack of a dynamic strategic plan to support the adoption of Industry 4.0 in the manufacturing industry |
| Chal. 7 | Environmental side effects | Huge use of automation in Industry 4.0 implementation may create serious environmental impacts |
| Chal. 8 | Complexity of reconfiguring the production pattern | Lack of capabilities to reconfigure the production pattern for the successful implementation of Industry 4.0 |
| Chal. 9 | Lack of skilled management team | Non-existence of skilled management team to execute the new and inventive Industry 4.0 business models |
| Chal. 10 | Complexity in integrating IT and OT | Intricacy of integrating information technology (IT) and operational technology (OT) |

**Table 5.9.** *The main challenges hindering the application of digital technologies in Industry 4.0 (Moktadir et al. 2018)*

| Number | Challenge | Difficult challenge | Easy challenge |
|---|---|---|---|
| Chal. 1 | Data uncertainty | – | 2 |
| Chal. 2 | High investments | 8 | – |
| Chal. 3 | Lack of technological infrastructure | 1-2-4 | – |
| Chal. 4 | Unstable connectivity among companies | – | – |
| Chal. 5 | Possible decrease in jobs | – | 1-3-6-8 |
| Chal. 6 | Lack of strategy for industry 4.0 | – | – |
| Chal. 7 | Environmental side effects | – | 4-5 |
| Chal. 8 | Complexity of reconfiguring the production pattern | 3-5-6-7 | – |
| Chal. 9 | Lack of skilled management team | – | – |
| Chal. 10 | Complexity of integrating IT and OT | – | 7 |

**Table 5.10.** *Ranking of challenges based on difficulty of application, as per the opinions from the groups of experts (Moktadir et al. 2018)*

To sum up, the procedure given in this example helps decision-makers identify and handle domains that constitute obstacles, before implementing industry 4.0 techniques in the supply chain of the leather industry. It also serves as a guide in implementing Industry 4.0 practices to improve process safety, occupational health and safety, and environmental protection. It should encourage stakeholders to develop digital technology 4.0 to facilitate treatment operations and to contribute to minimizing risks, especially those for personnel working with operational procedures. Finally, the major and inevitable implications of this process must be highly, namely:

– ensuring the development of Industry 4.0 IT infrastructure;

– defining strategic policies to reconfigure production patterns;

– convincing those in charge to adopt smart technologies.

## 5.15. Safety 4.0: metrics and performance indicators

The management of process industry installations involves, among other things, having pertinent performance indicators for risk control, first of all to assess the company's performance and, secondly, often in order to compare companies carrying out the same activity in a given sector. Nonetheless, the design and use of these indicators continue to pose a challenge both in scientific terms as well as in terms of industrial practice. Delatour et al. (2014) have listed the challenges that must be met in this regard.

| Number | Challenge | Weight |
|---|---|---|
| Chal. 1 | Data uncertainty | 0.1228 |
| Chal. 2 | High investments | 0.1225 |
| Chal. 3 | Lack of technological infrastructure | 0.2284 |
| Chal. 4 | Unstable connectivity among companies | 0.0780 |
| Chal. 5 | Possible decrease in jobs | 0.0415 |
| Chal. 6 | Lack of strategy for industry 4.0 | 0.0659 |
| Chal. 7 | Environmental side effects | 0.0369 |
| Chal. 8 | Complexity of reconfiguring the production pattern | 0.1900 |
| Chal. 9 | Lack of skilled management team | 0.0674 |
| Chal. 10 | Complexity of integrating IT and OT | 0.0465 |

**Table 5.11.** *Final result of the optimal average weights of the multi-criteria analysis of the challenges (Moktadir et al. 2018)*

In the absence of an international standard, Zwingelstein (2014) have suggested clearly defining the relationships between the different terms: the performance indicators, metrics and data. This is illustrated in Figure 5.11.

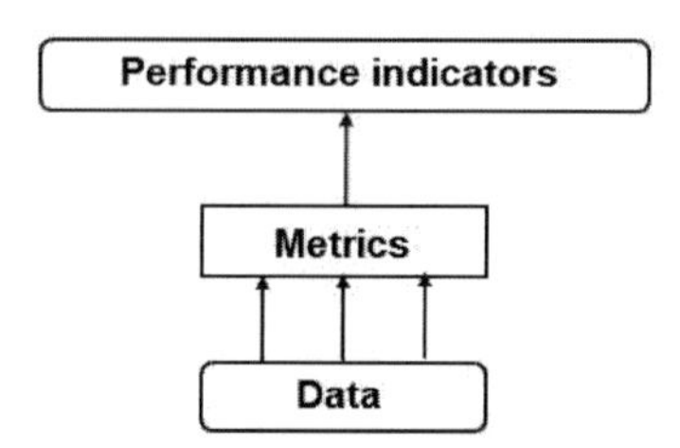


**Figure 5.11.** *Relationships between performance indicators, metrics and data (Zwingelstein 2014)*

Two groups of performance indicators are usually used:

– an indicator of impact (*lagging indicator*);

– an indicator of activity (*leading indicator*).

Figure 5.12 shows the respective positions of the lagging and leading indicators, upstream and downstream of the central critical event.

### 5.15.1. *Impact or lagging indicator*

A lagging indicator is an indicator that is said to be reactive a posteriori. It makes it possible to verify whether or not the desire result was achieved. It often measures the lack of performance or changes in performance. This indicator serves as an observation based on an existing fact, but does not indicate why the desired result was or was not achieved.

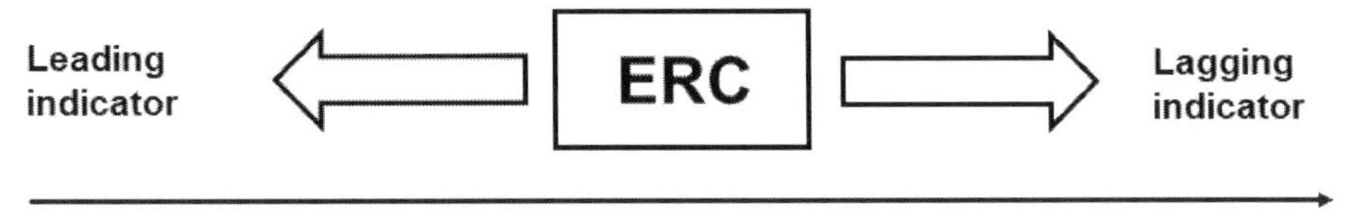


**Figure 5.12.** *Position of the lagging and leading indicators with respect to the central critical event*

### 5.15.2. *Activity or leading indicator*

A leading indicator makes it possible to identify whether or not a company has put into place proactive correction measures for unsatisfactory indicators. A leading indicator is related to the temporal and causal aspects of the operational and organizational systems in the company. Following a downstream analysis of causes, it makes it possible to improve the performance of systems and also to periodically verify that the priority actions are effective.

REMARK.– *A review of literature shows that most industries chiefly report the values of lagging indicators.*

*For example, many chemical companies publish and communicate, in the field of occupational health and safety, lagging indicators such as the frequency rate of accidents with stoppage of work, and the severity rates of temporary incapacities. It must be firmly reinforced that lagging indicators for work safety do not provide information on the control of major technological risks.*

*Thinking otherwise represents a major breach in effective risk management (Tazi 2014; Daniellou and Descazeaux 2019). It is, however, recognized that putting in place leading indicators is an effective risk management strategy.*

In the field of safety, Rogers et al. (2009) proposed a comparison of the function of both kinds of performance indicators. This is presented in Table 5.12.

### 5.15.3. *Some recommended examples of performance indicators for process safety*

There are many different and varied examples of performance indicators in industrial practice, and companies bear the responsibility of applying them. However, in activities that are considered hazardous to personnel safety, the safety of the plants or the environment, recommended and/or mandatory indicators are listed by different instances and standards (Zwingelstein 2014).

| Leading performance indicators | Lagging performance indicators |
|---|---|
| Low quantitative analysis | High quantitative analysis |
| Weak benchmark | Strong benchmark |
| Forecast of future outcomes | Strong data driving |
| Weak data driving | Controlling method for management level |
| Efficiency driven by measurement | Weak continual monitoring in the organization |
| Wide monitoring of all activities | Methods for investigation, analysis and record of weakness of health and safety management |
| Strong continual monitoring in the organization | Reactive key performance indicators to check achievement of goals and targets |
| Verification of consistency of health and safety related activities | – |
| Proactive key performance indicators to check achievement of goals and targets | – |

**Table 5.12.** *Functions of safety performance indicators (Rogers et al. 2009)*

### 5.15.3.1. *The case of a Seveso ICEP*

The French regulation resulting from the adoption of the European Seveso 1, 2 and 3 directives, implies, among other things, that those in charge of Seveso ICEP *(French: Installations Classées pour la Protection de l'Environnement, English: Installation Classified for Environmental Protection)* must put in place risk management measures (RMM). For example, the metrics for the leading indicators related to assessing the probability of occurrence and the severity of the consequences of a major accident have already been discussed in the regulatory matrix presented in Chapter 4 (section 4.2).

### 5.15.3.2. *Case of oil and chemical industry plants*

Several organizations, technical centers and professional federations have proposed standards and advisory documents for recommendations to identify and formulate leading and lagging indicators in the different targeted steps in the safety management system (CCPS 2010; HSE 2010; Mazri 2015; OSHA 2019; Rogers et al. 2009; Swuste et al. 2016; UICh 2017). Table 5.13 presents an example comparing the key domains of application of the performance indicators for process safety proposed, respectively, by CCPS, OSHA and HSE organizations.

| CCPS | OSHA | HSE |
|---|---|---|
| mechanical integrity | change management | inspection/maintenance |
| follow-up on action items | preventive maintenance | staff competence |
| change management | process hazard analysis | operational procedures |
| process safety training and competence | maintenance and mechanical integrity | instrumentation and alarms |
| safety culture | training | change management |
| operating | maintenance procedures subcontracting | communication |
| fatigue risk management | safety actions | work permit |
| – | – | plant design |
| – | – | emergency arrangements |

**Table 5.13.** *Key areas for the development of leading indicators of process safety*

For example, for the domain of "change management", the following list offers a few leading indicators:

– the design and definition of the modification system are correctly specified (HSE);

– the percentage of modifying actions undertaken during an appropriate risk assessment carried out prior to the change (HSE);

– the percentage of modifying actions undertaken where the changes and results have been documented (HSE);

– the percentage of modifying actions undertaken for which authorization was given prior to implementation (HSE and CCPS);

– the percentage of modifying actions for which post-modification checks were carried out (HSE);

– the number of modifications that were delayed (OSHA);

– the number of modifications that were approved (OSHA);

– the number of ongoing, open modifications (OSHA);

– the number of modifications carried out each month (OSHA);

– the percentage of chosen modifications that satisfied all aspects of the change management procedure (CCPS);

– the percentage of restarts, following the modifications to the plant, for which no safety problem related to the modification was encountered during the reactivation (CCPS).

Although section 5.15.3 was largely dedicated to leading indicators, it must be noted that the reciprocal interaction between leading and lagging indicators often helps in interpreting and improving the creation of the safety management system. With regard to this, a very useful, non-prescriptive orientation document on safety performance indicators for prevention, preparation and intervention with respect to chemical accidents was published by the Organization for Economic Cooperation and Development (OECD) in 2008 (OECD 2008). It makes it possible to define the levels of application and use of performance indicators by operators of a process, site personnel and staff in the company, populations beyond the site, public authorities and all other stakeholders. In terms of the order of magnitude, this document lists about 100 lagging indicators and 220 leading indicators of safety in the chemical industry.

### 5.15.4. *Examples of the application of safety performance indicators*

Two industrial examples are presented here to illustrate how performance indicators contribute to better risk management.

#### 5.15.4.1. *Indicators and safer manufacturing*

There are often many potential approaches to manufacturing a chemical product. A global evaluation, a priori, of the approach from the design phase in research and development could make it possible to envisage modifications to one or more steps, or even propose an alternative process to try and define a safer process, especially in terms of occupational health and safety of the concerned operators. To recall, this is one of the principles to help control risks proposed by Kletz (1991, 1998) and CCPS (1996).

The example here concerns the manufacture of phenol. Four phenol manufacturing methods were studied as follows:

– the Hock phenol synthesis process, using the Cumene pathway (identified by CO), has three exothermic steps:

- alkylation of benzene into cumene by propylene,
- oxidation of cumene into cumene hydroperoxide,
- scission of cumene hydroperoxide into phenol and acetone;

– the Dow process of toluene oxidation (denoted by TO) in two steps:

- oxidation of toluene to benzoic acid,

- oxidative decarboxylation to obtain phenol;

– direct oxidation of benzene in liquid phase using hydrogen peroxide (denoted by DBL);

– direct catalytic oxidation of benzene in gaseous phase (denoted by DBG).

Hassim et al. (2010) proposed using a performance indicator from the R&D design stage, $I_{IOHI}$, based only on the chemical properties and reaction conditions available in this early stage. The method, named the "Inherent Occupational Health Index" (IOHI), comprises two indicators:

– $I_{PPH}$, the leading indicator, which represents the possibility of workers being exposed to chemical products;

– $I_{HH}$, the lagging indicator, which characterizes the health impact of the hazard due to exposure.

The performance indicator is calculated for each potential cumene manufacturing process using the relation:

$$I_{IOHI} = I_{PPH} + I_{HH} \qquad [5.1]$$

The leading indicator, $I_{PPH}$, consists of six weighted metrics: the process mode, the state of the components, their volatility, the pressure, the temperature and resistance to corrosion. The lagging indicator, $I_{HH}$, consists of data from the limiting exposure values and the nature of the earlier scores for the risk phrases (R phrase). The different contributions and weights of the data are detailed by Hassim et al. (2010).

Figure 5.13 illustrates an example of the comparison of results for four processes in the additive, moderate and worst-case scenarios.

The ranking of the additive scenario, obtained through the sum of the metrics for all the steps, shows that the cumene oxidation process (denoted by CO) is likely to be the most hazardous, because of the presence of three steps. The ranking continues with the two-step toluene oxidation process (denoted by TO). The ranking of the processes using the direct oxidation of benzene (DBL and DBG), which have only one step, are separated by the values of the respective reaction temperatures. The moderate scenario, estimated by overcoming the influence of the number of steps in the processes, shows here that the CO and TO processes become the most favorable alternatives. Finally, the worst-case scenario, which only looks at the weights of the most detrimental metrics, indicates that the direct oxidation of benzene in its gaseous phase (DBG) would be the most hazardous process.

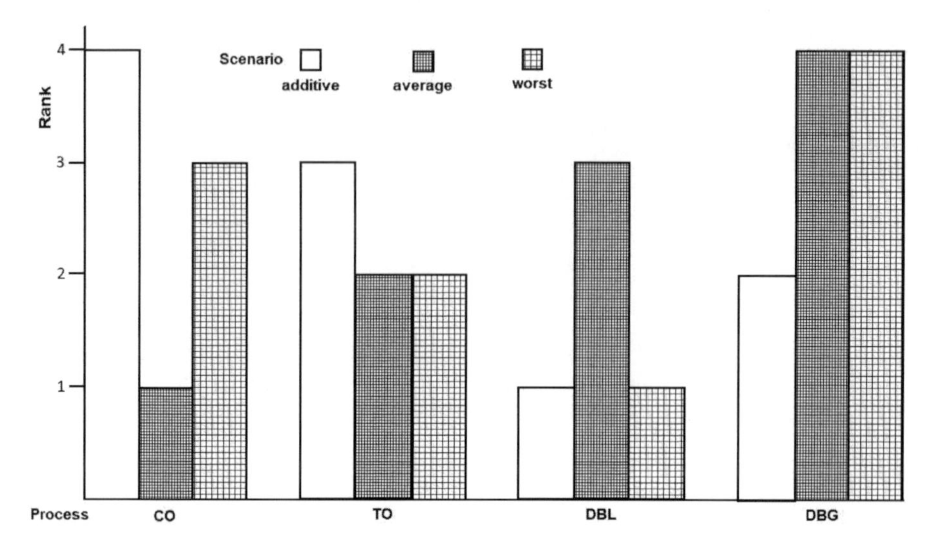


**Figure 5.13.** *Comparative rating scale for the danger levels of four Cumene manufacturing processes (level 4 being the most dangerous) (Hassim et al. 2010)*

In conclusion, the IOHI method could make it possible to compare the relative values of the four procedures being studied. A sensitivity study on the influence of the different metrics must complete this analysis.

#### 5.15.4.2. *Performance indicators in a large chemical group*

This example shares the experiment, carried out by the BAYER group Leverkusen (D), in managing safety by implementing performance indicators (Klein and Viard 2013). A large program to develop overall process safety management, comprising various robust operational concepts, reliable prevention and effective attenuation, was reinforced by four performance indicators, of which three were leading indicators and one was a lagging indicator.

##### 5.15.4.2.1. The "risk analysis leading indicator"

The first leading indicator, called the "risk analysis performance", was related to the development of the concept of safety, both in analyzing process risks based on a HAZOP methodology adapted by Bayer and the implementation of defined actions and measurements, either before the launch of a new unit, or at an appropriate or desired time for units that were already running. The objective was to guarantee and demonstrate that the safety concepts for the plants were updated and that no installation was being run without an updated safety concept being implemented. This indicator is computed as the sum of the number of delayed risk analyses and the number of delayed safety studies before start-up and/or the receipt of modification of use, divided by twice the number of existing risk analyses. The value is then multiplied by 100 to yield a percentage.

The REX showed that the “risk analysis” indicator has proved to be an excellent tool for the international community of experts on the safety and processes in the Bayer group, to initiate discussions on the harmonizing of process safety management at the level of all of the group’s factories and installations worldwide. The quarterly publication of the value of this indicator and the analysis of how it has evolved makes it possible, for example, to formulate such questions:

– What would be the optimal number of risk analyses to carry out in an installation?

– How to share information relative to process safety at a global level in the Bayer group?

– How to exchange information on risk analyses in order to ensure a greater and constant level of safety in all processes and across all factories?

– What is the most efficient approach for preparing periodic re-validations of risk analyses?

– What is the desired global approach to management of change?

– How can REX on current work be organized to plan for improved safety conditions in the future?

#### 5.15.4.2.2. The leading indicator “inspections and validation tests”

The second leading indicator is called “inspections and validation tests”. Mechanical integrity is a passive measure that is essential for preventing the discharge of substances in a chemical process. Carrying out inspections within deadlines is essential to guarantee the integrity of equipment. The indicator includes metrics related to inspections carried out based on the time and equipment and piping hazards in the process, as well as the safety valves. It also includes data from regular control tests and validations, which are carried out to guarantee that undetected, potentially hazardous failures are identified and that the safety locking features function reliably on demand. This indicator is defined as the ratio, in percentage, of the sum of the number of regulatory and internal inspections not performed, and the number of proof tests not completed on time, to the total number of inspections and control tests to be carried out in a calendar year.

#### 5.15.4.2.3. The leading indicator “availability of the safety concept”

The third leading indicator expresses a measurement of the availability of a safety concept. The learning gleaned from various incidents that have occurred in the chemical industry show that the bypassing and deactivation of safety lock systems were contributing factors to the consequences that were seen. This indicator thus contributes to accounting for the number of safety locks present in safety instrumentation systems (SIS), or equivalent systems, which were bypassed unintentionally. This indicator is computed from the ratio, in percentage, of the number of bypassed safety locks in the quarter of the year under study to the total number of existing safety locks. It must be noted that the number of locking mechanisms deactivated deliberately and temporarily during startup, shutdown or control and validation tests are not taken into account in this indicator, since these procedures are governed by the management of authorizations and permits.

The main advantage of this indicator is clearly to increase and maintain attention and awareness among the personnel of the company. That concerns the operations for the bypassing of safety interlocks and the requirement to regularly check the status of the safety interlocks in the plant.

##### 5.15.4.2.4. The lagging indicator "primary failure in integrity of the container"

The lagging indicator of the loss of integrity of a container is defined based on the definition proposed for process safety incidents by the International Council of Chemical Association (ICCA 2017). A database was set up, counting the number of incidents of loss of integrity of a container and studying the causal factors in these incidents. This indicator, integrated into the monthly safety report, is defined, tracked and communicated using the same tools and presentations used in the field of occupation health and safety for the indicators of the frequency rates of work accidents.

These two examples illustrate the potential for using safety performance indicators as elements in enhancing risk management.

# 6

# Intensification and Inherent Safety: Myth or Reality?

This chapter first reviews a few essential elements of the process intensification in the context of industries of the future. It then describes a few classic examples of process intensification. Finally, the principal objective of this chapter is to try and use specific examples to try and take stock of Safety 4.0, attempting to specify its limitations in order to separate myth from reality when it comes to the effectiveness of Safety 4.0.

## 6.1. A review of essential elements in process intensification

In the *European Roadmap for Process Intensification*, this phenomenon is defined as a set of radically innovative principles (paradigm shift) for process design and technologies that bring in improvements by a factor greater than 2, in terms of efficiency, costs (investment and operation), quality, waste and safety (Devries 2007). Gourdon (2016) considers that the claim of a paradigm shift in the definition corresponds more to innovation, which can be termed "disruptions", rather than incremental changes. Gourdon et al. (2018) summarized the objective of this definition through the simple statement that process intensification consists of using much *less* to produce *more* and *better*. The *less* is related to investment, space, time, primary material, energy, stocks, etc., while the *more* is in terms of factors or orders of magnitude.

Process intensification has been the subject of much study and many industrial and university developments (Charpentier 2002, 2005, 2007, 2016; Stankiewicz and Moulijn 2003; Commenge et al. 2005; Becht et al. 2007, 2009; Hessel 2009; Hessel et al. 2011, 2013; Falk et al. 2019). According to Gourdon (2016), the fundamental principles of the concept of intensification are as follows:

– optimizing chemical transformation by being closer to intra- and inter-molecular events;

– ensuring each molecule is treated identically;

– reducing the transfer limits on mass and heat;

– searching for synergy effects between phenomena and/or operations.

The implementation of these principles involves a multi-physical and multi-scale approach built on four pillars:

– the structure (spatial level);

– thermodynamics (the operating window and energy contribution);

– synergy (multifunctionality and/or hybridization);

– time (temporal scale).

The transposition from batch to continuous process, the miniaturization of equipment on a spatial scale using micro- or milli-technologies, energy input, multi-functionality, hybridization of technologies, and transient, cyclic or periodic regimes with a time scale are examples of these elements. Figure 6.1 shows a classification of the tools for intensification, distinguishing between equipment (hard tools) and methods and techniques (soft tools). This classification thus makes it possible to associate a given technology with a specific intensification goal.

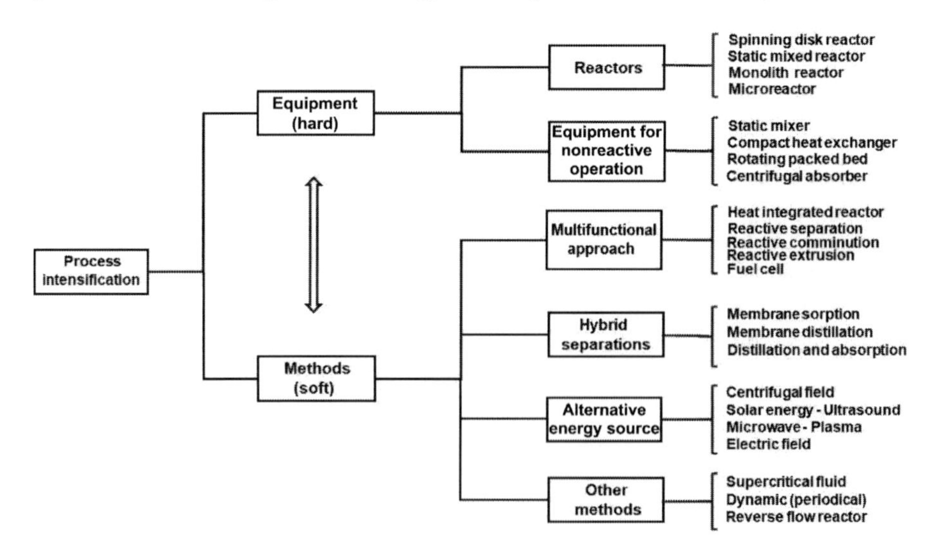


**Figure 6.1.** *Classification of tools and intensification (Devries 2007)*

There are other approaches, such as intensification from the application of different rules like the integration of unit operations, the integration of functions, the integration of phenomena and targeted intensification of phenomena within a given operation (Gourdon 2016). On the other hand, the approach based on the analysis of the characteristic times consists of listing elementary phenomena by associating their characteristic time scales and then comparing the technologies based on their respective performances, also expressed in characteristic times (Commenge et al. 2004, 2005; Commenge and Falk 2014; Florent et al. 2013).

This brief review can be concluded with an analogy with "Triple-P" sustainable organization philosophy. The triangular planet (Earth)–population–profit diagram in Figure 6.2 illustrates this analogy with the advantages of process intensification.

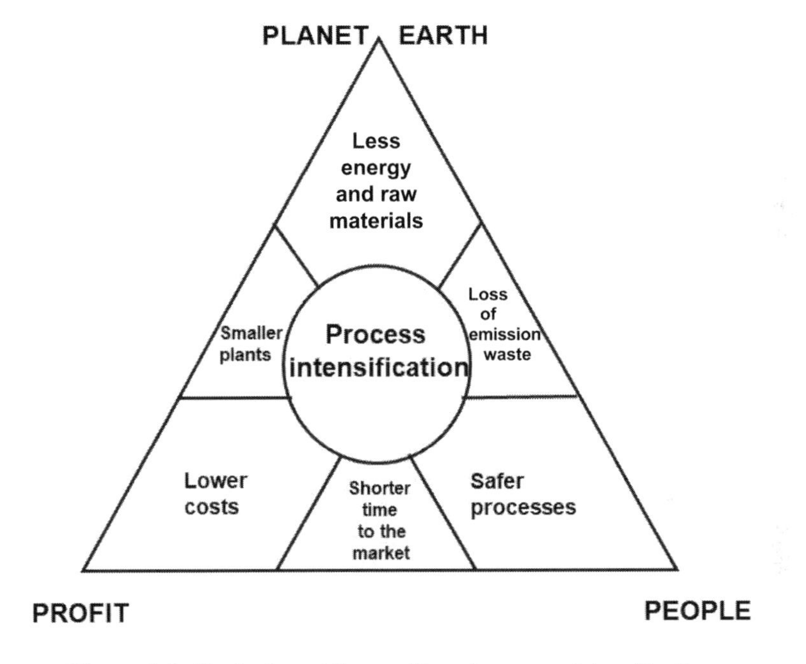


**Figure 6.2.** *Illustration of the position of process intensification within a sustainable organization (Devries 2007)*

## 6.2. Some examples of process intensification

### 6.2.1. *The reduction principle in support of the risk management*

The term *process intensification* first appeared in the 1980s, while the concept of inherently safer design (ISD) was introduced by Trevor Kletz, the Senior Councillor for Process Safety for the company ICI (UK), from the 1970s, to define about 10 principles to support the industrial risk management. These principles are now largely widespread in literature (Englund 1990; Kletz 1991, 1998; CCPS 1996).

The reduction principle, now sometimes called the intensification principle, proposes, wherever possible, to reduce the volumes and flows of dangerous substances to the minimum levels compatible with the proper functioning of the unit (Hendershot 2000).

The manufacture of nitroglycerine is a classic example of the application of the reduction principle (Kletz 1991). Initially, the exothermic reaction of glycerine with concentrated nitric acid in the presence of sulfuric acid was carried out in a batch reactor containing 1 ton of load. Following the redesign, the product is manufactured in a continuously stirred flow tank reactor (CSTR), with a 1 kg liquid.

Another solution that applies the reduction principle is the in situ production and consumption (without the intermediary storage) of the reactant or hazardous intermediary. For example, during the polymerization in suspension of vinyl composites in a batch reactor, diisopropylperoxycarbonate (IPP) is used as the initiator because of its short half-life and its good polymerization properties.

On the other hand, storing IPP is delicate, difficult and potentially hazardous, and it must be stored between −10°C and −20°C. The on-demand in situ generation can be carried out through a reaction with oxygenated water diluted with isopropylchloroformate in the aqueous phase of the suspension just before initiating polymerization (Englund 1990). The process is then inherently safer, despite the presence of the two new reactants, although it presents a lower risk potential than the IPP storage.

Another illustration consists of transposing a reaction in a batch reactor to a reaction in a semi-batch reactor (Englund 1982a,1982b, 1990, 1991, 1995; Gowland 1996). The preparation of a styrene-butadiene latex can be used in a batch reactor by initially introducing loads of monomers of styrene and butadiene on the one hand, and an emulsifying aqueous solution (sodium lauryl sulfate) and an initiator (sodium persulfate) on the other hand. In the configuration in Figure 6.3 showing the functioning of a batch reactor, the total initial load in the reactants presents a higher risk potential (inflammability, thermal runaway, etc.).

An inherently safer situation can be obtained through reduction with a semi-batch reactor by progressively slowing adding, on the one hand, the mass flow of the two monomers and, on the other hand, the charge of the aqueous solution of emulsifier and initiator (Figure 6.4). The quantity of the monomer reactants present in the reactor is consequently minimized, the reaction temperature is better controlled and finally the productivity and the quality of the latex obtained are better than in the batch process, because of the adequate elimination of the exothermic nature of the reaction, with the additional presence of a reflux condenser.

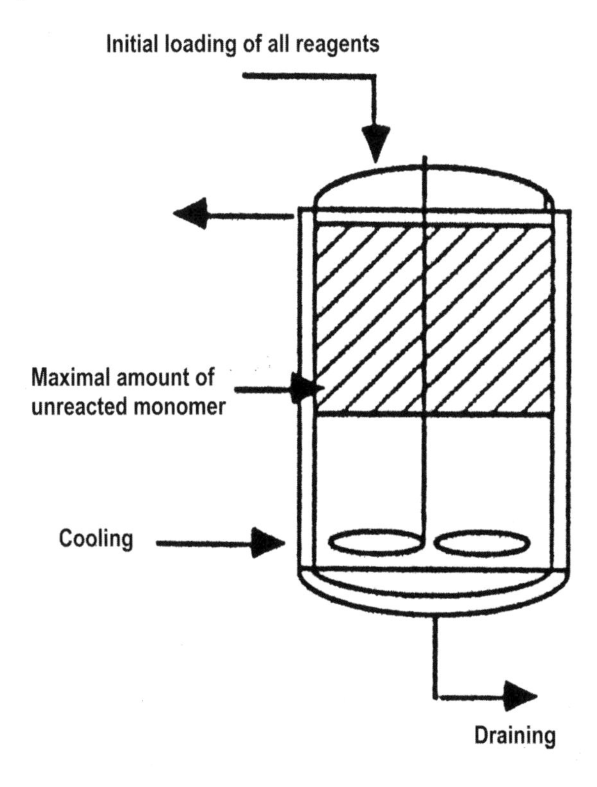


**Figure 6.3.** *Functioning in a batch reactor (Englund 1990)*

The use of "mini-units" (miniaturization process), whose geometric scale is located between the pilot scale and the "microtechnical" scale, offers another opportunity for applying the reduction principle and thus contributing to the emergence of inherently safer processes. In an article that summarized the prospects and potential of "mini-units", Behr et al. (2000) presented examples to illustrate this, either through integrating the "mini-unit" in the process or through production via the "mini-unit". The ozonation treatment of water is cited as an example of integration. The manufacture of fatty alcohol ethoxylates is depicted for a production process using two "mini-units", one producing the ethylene oxide required for the other to carry out the ethoxylation. Swan (2000) has also highlighted the advantage of implementing manufacturing microtechnology to apply the reduction principle, for example, during the production of phosgene and hydrogen cyanide.

Several other examples are described in *Process Plants: A Handbook for Inherently Safer Design* (Kletz 1998).

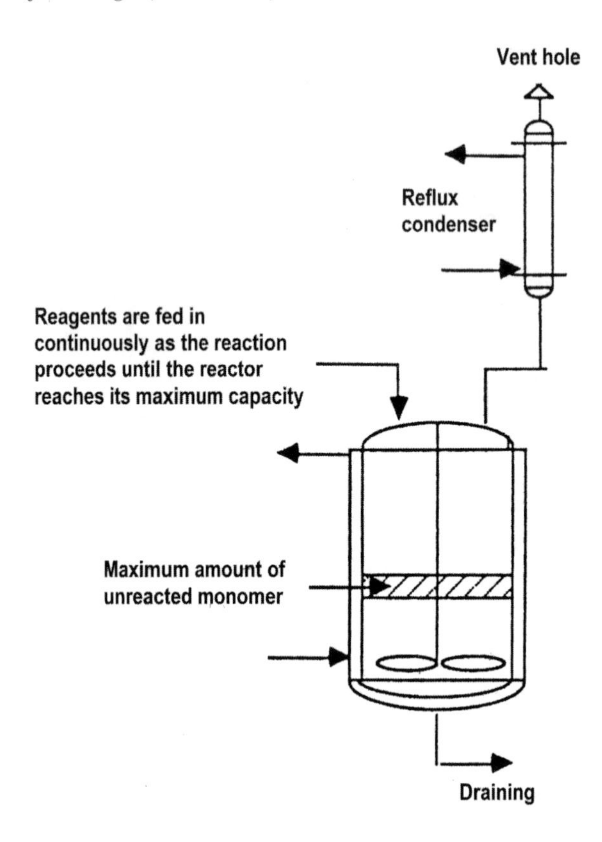


**Figure 6.4.** *Functioning of a semi-batch reactor (Englund 1990)*

## 6.2.2. *Areas of interest for using microstructured reactors*

Krummradt et al. (2000) showed that it was possible, for instance, studying the same synthesis of organometallics in fine chemistry in many types of reactors of different scales to reduce the operation time of 6 h in a semi-batch mixed industrial reactor of 6 $m^3$ to less than 10 seconds in a continuously operating microreactor. Initially, these authors had actually extrapolated, with a factor of 23 with respect to the diameter, the results obtained from a semi-batch stirred reactor with a volume of 0.5 L to a batch of 6 $m^3$ industrial reactor.

Given the highly exothermic reaction of the chemical synthesis, monitoring the internal temperature of the reactive volume is an important parameter. A good yield and good selectivity of the final product involve the implementation of a cryogenic process. The required temperature has proven to be difficult to meet on an industrial scale. This then leads to the formation of secondary products. Consequently, Krummradt et al. (2000) tested equipment that was made in the Institute of Microtechniques at the University of Mayence (Germany) and that was composed of 10 static micro-mixers, arranged in parallel. Each of these contains 32 channels that have oblong right-sections (220 mm long; 40 $\mu$m wide). However, fouling due to solid deposits disrupted the functioning of this system. Ultimately, the solution that was chosen was installing five mini-mixers, characterized by an $A/V$ ratio ($A$ - heat exchange area/$V$ reactor volume) of 4,000 $m^{-1}$. This installation was successfully operated at Merck for 5 years, until the end of life of the fine chemical product (Hessel and Lowe 2005).

Lomel et al. (2006) explained these experimental results by examining the reduction in the operational timings, based on the intensity of the mixture and the characteristic time, $t_{tr}$, for the heat transfer due to the change in scale in the configurations under study. By solving the balance equations for mass and heat in the isothermal conditions of a semi-batch reactor, a continuously stirred reactor, a plug flow mini-reactor and a plug flow micro-reactor, Lomel et al. (2006) were able to show, for a given fixed-conversion example, the effect of geometric structuring on the process performances. Table 6.1 summarizes the predictive values obtained.

| Reactor | Diameter (m) | U (IS) | A/V ($m^{-1}$) | Characteristic time of heat transfer, $t_{tr}$ | Operation characteristic time |
|---|---|---|---|---|---|
| Semi batch | 2 | 200 | 2.4 | 2.5 h | 7 h |
| CSTR | 2 | 200 | 2.4 | 2.3 h | 5.6 h |
| Mini plug flow | 0.001 | 2,200 | 4,000 | $5\times10^{-1}$ s | 1.5 s |
| Micro plug flow | 0.0001 | 22,000 | 40,000 | $5\times10^{-3}$ s | $1.5\times10^{-2}$ s |

**Table 6.1.** *Estimation of the operating time based on the type of reactor (Lomel et al. 2006)*

Table 6.1 shows that the configurations of the mini-reactor piston and micro-reactor piston are the most efficient in terms of operating times. However, the micro-reactor piston offers a very short operating time, which is not realistic in practice for the reaction regime being considered.

Before moving away from this example, it must be observed that the major part of the work on intensification, in the fine chemical and pharmaceutical domain (Hessel and Lowe 2005), often consists of experimentally testing the relevance of using a structured reactor for a given complex reaction system and the conclusions often remain empirical. The procedure suggested in this example makes it possible to understand whether a micro- or mini-reactor will actually be useful in a given situation. The understanding and demonstration of reaction phenomena on a laboratory scale, and the potential for the implementation in parallel of such continuously functioning reactors, demonstrates the advantage they offer. In particular, it was possible to transition from a conventional, discontinuous batch process to a continuous process by comparing the performance of both equipment in the process.

### 6.2.3. *Transposition of an exothermic reaction in an intensified, continuous heat exchanger*

The implementation and control of exothermic reactions that can provoke thermal runaway are still often carried out in discontinuous batch reactors.

In spite of their flexibility and polyvalence, especially in the field of fine chemical products or pharmaceuticals, discontinuous batch reactors pose an important safety problem because of the limitations seen in the conditions for removing reaction energy. An alternative to using discontinuous reactors is to transpose exothermic reactions into continuous plug flow reactors, which offer better control over heat transfers through intensification.

The Open Plate Reactor (OPR), developed by Alfa Laval Vicarb, is a new concept for an intensified reactor that is multi-functional, continuous as well as smaller in size. It combines the functionalities of a chemical synthesis reactor and those of a heat exchanger. It offers a volume that is reduced to the order of a liter and allows a continuous fluid flow of reactants, products, catalysts and utilities within the reactor. The OPR has been designed based on a modular block structure with a plate exchanger. Figure 6.5 depicts the scheme for the principle of the internal, sequential structure of the different plates (Benaissa et al. 2008b).

The internal structure of the OPR is made up of a series of plates of various natures:

– the reaction plate (RP), comprising an assembly of inserts, in which the reactants flow, is made of PEEK, a thermoplastic material that is corrosion resistant and can tolerate high temperatures;

– the sandwich plates (SP), in stainless steel, placed on either side of the RP;

– the UF plates that correspond to the zone of circulation of the utility fluids;

– the transition plates (TP), adjacent to the UF plates, which ensure thermal insulation between two blocks.

Benaissa et al. (2005, 2008a) studied the transposition of the esterification reaction of propionic anhydride by 2-butanol, leading to the formation of butyl propionate and propionic acid. The process generally used for this synthesis is the semi-batch type. The procedure consists of first introducing the alcohol and catalyst (generally sulfuric acid) into the reactor chamber. After achieving thermal equilibrium at the operational temperature (30–80°C), propionic anhydride is added in a stoichiometric quantity to the reaction mixture. The reaction is carried out at atmospheric pressure. Depending on the size and cooling capacities of the reactor, the addition time for the second reactant can vary from a few seconds to tens of minutes. In terms of safety, this reaction is a particularly interesting case study of the procedure to assess and predict the risk of thermal runaway. Benaissa (2007) summarizes the following elements of this process:

– the reaction is carried out in a homogeneous medium;

– the reaction is moderately exothermic;

– there is an elaborate kinetic model;

– the reaction rate is of a second-order reaction, in the absence of a strong acid (it is a first-order reaction vis-a-vis each reactant) and self-catalyzing in the presence of sulfuric acid;

– thermodynamic data for the reaction is accessible;

– the physicochemical data for the reactants and products are available.

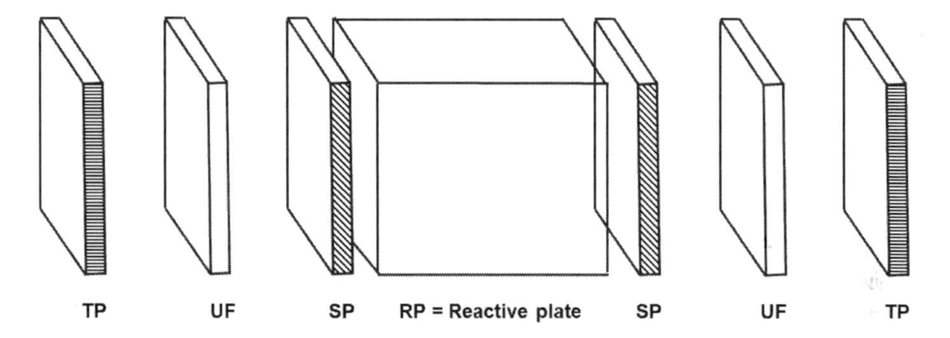


**Figure 6.5.** *Scheme of the principle for the internal structure of the OPR plates*

Having experimentally characterized the thermal and kinetic stability of the reaction using calorimetrical analyses, a simulation program for the dynamic behavior of the reaction for a discontinuous/semi-continuous process was established from the solutions to the mass and energy balance equations, and from the kinetic model. The experimental validation versus simulation made it possible to determine operating conditions for the reaction in a semi-continuous reactor. The synthesis was then transposed to and studied in the OPR. The details on the procedure that was successfully carried out are presented by Benaissa et al. (2005, 2008a, 2008b) and Benaissa (2007). It is interesting to study the results that these authors obtained to understand the criticality of the dynamic behavior of a scenario involving a deviation from the process, such as the utility fluid's flow in the OPR stopping. Drawing inspiration from the conventional model of the three parameters representation diagram, which describes the state of evolution of a batch reactor, the authors evaluated the corrected adiabatic rise in temperature and the time taken to reach the maximum rate for a decomposition reaction in the case of various hypotheses for thermal inertia in the OPR.

Computing the characteristic time for thermal diffusion in the OPR made it possible to show that the hypothesis in which the reaction energy was instantly dissipated in the steel and PEEK SP is acceptable. The result was that taking into account these corresponding exchanger-reaction masses led to a decrease in the maximum temperature reached in the case of a deviation, by an order of 20–60°C, with respect to the adiabatic case considered in a batch reactor. The maximum time taken to reach the maximum speed for the decomposition reaction also saw considerable increase from 18 min to 3 h (Benaissa 2007). Finally, the proposed methodology was applied in the OPR to the rapid and strongly exothermic reaction of the oxidation of sodium thiosulfate through oxygen peroxide to form sodium trithionate and sodium sulfate.

### 6.2.4. *Pilot demonstration of IMPULSE for the production of sulfur trioxide through the oxidation of sulfur dioxide by air*

The integrated European project IMPULSE (Integrated Multiscale Process Units with Locally Structured Elements) was designed in 2005. It brought together 20 university and industrial partners from seven countries, and it was a flagship project in the "Chemistry" sector for the 6th Framework Programme for Research and Technological Development (FPRTD) in the European Union (Bayer et al. 2005). The IMPULSE project was based on the implementation of two principles:

– process intensification to significantly reduce the size of a process, while preserving production capacity;

– miniaturization to do better, smaller, more compactly and more precisely.

The integrated IMPULSE project aimed to achieve this by equipping chemical production systems with micro-reactors. The goal was to arrive at multi-scale equipment that performed better in terms of quality, efficiency and safety. However, the integration of these microtechnologies into more "traditional" macro-processes posed new problems and constraints. In particular, even if these micro-reactors are now often considered to be inherently safer, as they contribute to a reduction in the volume of chemical products involved, at any time in the process, there was as yet no formalized and recognized method to evaluate the safety levels of these new installations. Further, these new difficulties, related to hydrodynamics, mass and heat transfer, fouling in micro-channels and interconnectivity between macro- and micro-equipment, were not covered by conventional risk analysis methods. Consequently, in the framework of this project, there was a sub-group whose specific mission was to develop new methods and tools to analyze, assess and control accident risks introduced by the intensification of chemical processes. The example reported here illustrates one of the responses from the deliverables generated by the integrated IMPULSE project. This example looks at an installation, at a pilot stage, demonstrating the production of sulfur trioxide through the oxidation of sulfur dioxide through air.

### 6.2.4.1. *Conventional process*

The traditional process for the production of sulfur trioxide through the oxidation of sulfur dioxide by oxygen is well established at the industrial scale. A production unit generally comprises the following four sections:

– combustion of sulfur;

– catalytic conversion of sulfur dioxide to sulfur trioxide;

– absorption of sulfur trioxide into sulfuric acid;

– cooling of the acid.

Most units function at oxygen pressure that is greater than atmospheric pressure.

### 6.2.4.2. *New process in the pilot plant*

The new process results from the development of a laboratory reactor, in its pilot plant, for the oxidation of sulfur dioxide to sulfur trioxide, by replacing oxygen with air. This prototype for the reactor was also implemented in pilot installations in the operating company Procter and Gamble, Brussels, in cooperation with the micro-reactor manufacturer, Institut für Mikroverfahrenstechnik (IMVT) in the Karsruhe Institute of Technology (KIT) in Germany. The reactor is part of a process intensification assembly, in which the sulfur trioxide produced reacts directly with organic alcohols in a connected falling film reactor. The diagram depicting the principle behind the pilot reactor for the demonstration is given in Figure 6.6.

The oxidation reaction of sulfur dioxide, which is strongly exothermic, is a reversible reaction, carried out in an excess of air in gaseous phase, under an absolute pressure of between 1 and 5 bars. The reactor being studied contains two sections, operating in different temperature domains, and which are separated by an adiabatic intermediate zone. Each section is made up of a stack of structured plates, welded by diffusion welding. The length of the interstitial micro-channels is 8 cm. The catalyst is deposed on the internal surface of the micro-channels. The upper section, operating at between 500 and 530°C, promotes the conversion of sulfur dioxide at equilibrium. The lower section, operating at a lower temperature of between 400 and 420°C, promotes the shift in equilibrium toward sulfur trioxide to end the conversion. The temperatures in the two sections are controlled by two independent cooling circuits, working at cross-current with respect to the direction of the reactant gaseous flow, in order to satisfy the heat balance equations (Klais et al. 2010b).

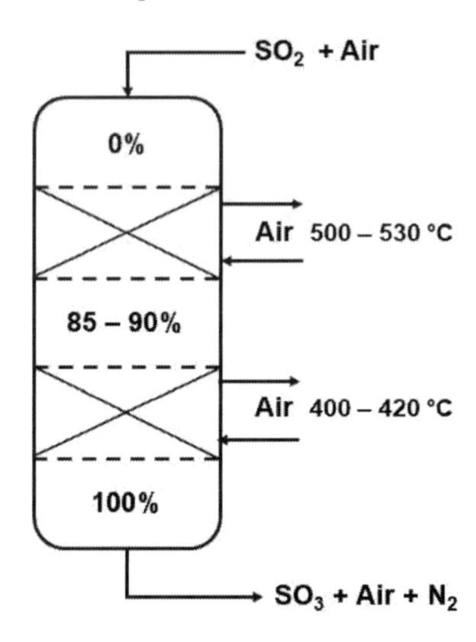


**Figure 6.6.** *Scheme depicting the principle for the pilot reactor for the demonstration (Klais et al. 2010a)*

The authors analyzed and validated the safety of this process using the generic canvas of the guidebook for analyzing safety and health in multi-scale process intensification, which will be examined further on in this chapter, in section 6.5 (Klais et al. 2010a).

### 6.2.5. *Synthesis of ionic liquids by alkylation in a microstructured reactor*

The synthesis of ionic liquids via alkylation is generally carried out in batch or semi-batch conditions in stirred tank reactors. The reaction is highly exothermic and rapid. Managing the heat is a key point in conventional production, in conjunction with prolonged dosing time, in order to keep the reaction temperature of the discontinuous process under control.

Based on laboratory studies, an intensification unit to demonstrate the production of 1-ethyl-3-methylimidazolium [EMIM][EtSO4] by alkylation of methylimidazole [MIM] with diethyl sulfate [DES], without a solvent, was designed and implemented at the Institut für Technische und Makromolekulare Chemie, University RWTH

Aachen (Germany). The two reactants are liquid and miscible at room temperature, such that the reaction can be carried out without a solvent. The operating capacity is from 20 to 100 kg of product per day. The demonstrator platform operates under a total pressure of 40 bars in the liquid phase. Each reactant is introduced via a dosing syringe pump into a static mixer arranged upstream of the microreactor. The reactive mixture that is obtained circulates successively through the microreactor (reference type HPMR-2TZ-90 ml-V2) and in an additional cascade of microreactors with capillary ring channels placed in series. This arrangement makes it possible to improve the conversion rate. Finally, depending on the output concentration of the product measured in the mixture, a bypass valve directs the product flow toward the receiving reservoir or the waste reservoir. The details of the installation are described by Klais et al. (2010a). The whole assembly makes up a compact unit enclosed in an airtight rack, with a volume of 4 $m^3$, with transparent screens on the sides. Figure 6.7 shows a photograph of the installation (Matlosz 2009). The operational safety of this demonstration unit was considered satisfactory.

### 6.2.6. *Developing an intensified process for the industrial synthesis of methanol from carbon dioxide*

Methanol is a chemical intermediate that can be used to produce fuels such as dimethylether and various chemical products. At present, methanol is synthesized on an industrial scale from the reactants hydrogen and carbon dioxide, and this is carried out in fixed bed catalytic multitube reactors with zinc oxide and copper catalytic particles, operating at high pressure (50–80 bar) and at temperatures between 200°C and 300°C. Thermodynamic studies of gaseous mixtures containing carbon monoxide, carbon dioxide and hydrogen show that the production of methanol is promoted by high pressures and low temperatures (150–200°C). On the contrary, high temperatures are more conducive to the kinetics of the synthesis. A compromise must be made between the kinetics and thermodynamics to optimize the desired reaction.

This example aims to take a comparative look at the impact that the arrangement and structuring of the solid catalyst have on the performances of conventional fixed particle bed tubular reactors, and a monolithic microstructured reactor (Arab et al. 2014). The conventional reactor contains tubes that are 2 cm in diameter, 0.5–8 m in length, containing spherical catalytic particles from 1 to 3 mm in diameter. The microstructured monolith reactor is made up of cylindrical tubes that contain several monolithic channels, with a square cross-section of 1–2 mm along the edge, the same length as the tubes of the conventional reactor, with a catalyst thickness of 0.1–0.6 mm. The operating conditions studied are as follows:

- input temperature of the reactants: 240°–260°C;
- temperature of the tube walls: 250°–270°C;

– pressure: 80 bar;

– composition of the reactants: ($CO_2$, $H_2$, $N_2$)-(24, 72, 4% molar);

– weight hour space velocity (WHSV): 2–10 $h^{-1}$.

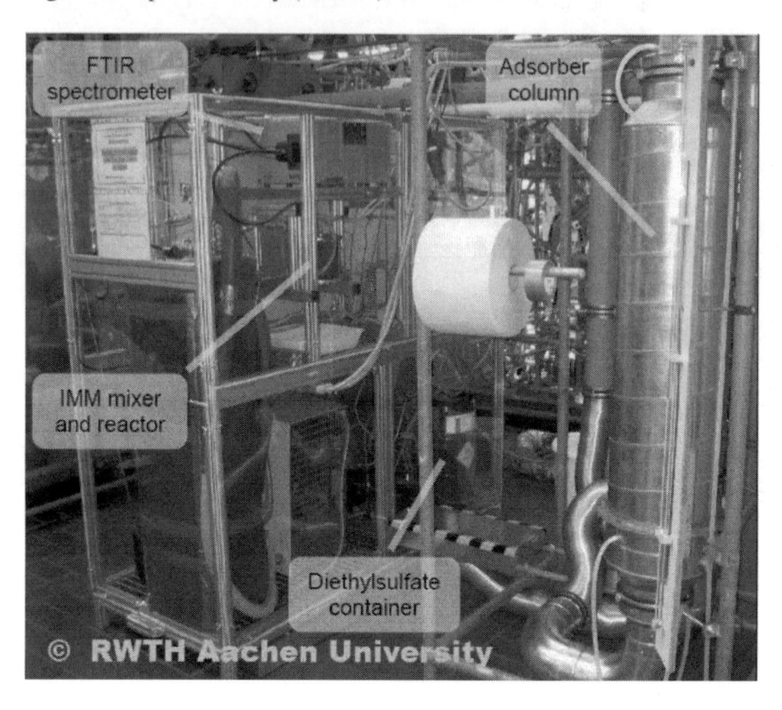


**Figure 6.7.** *Photograph of the rack containing the compact micro-unit (Matlosz 2009)*

Arab et al. (2014) studied the behaviors of the two reactors using a heterogeneous model, adapted for each configuration, based on their geometric parameters and the mass transfer and heat transfer parameters. Taking into account the risk of the emergence of hot points that would cause the deactivation of the catalyst, the temperature gradient in the tube had to be limited to 10°C. Both technologies were compared by examining the methanol yield by considering the same catalyst mass in each reactor, depending on the weight hour space velocity (WHSV). For the WHSV values that are currently used to synthesize methanol on an industrial scale, using fixed-bed reactor technology remains the most appropriate. This reactor is easy to manufacture and operate, and it is widely used. On the other hand, the use of monolithic reactors for the same WHSV values is not appropriate, as they are more expensive to manufacture and lead to performances similar to those of fixed bed reactors. For high WHSV values, the fixed bed reactors are penalized by high pressure drops. Monolithic reactors are more efficient by operating at higher spatial velocities, with improved performance and negligible pressure drop. The development of reactor-separator technology, to eliminate products, is a strategy that would help not only in intensifying reactors, but in intensifying the process in its entirety. This technology could be used in these conditions to synthesize methanol. However, a more in-depth technical and economic study of monolithic reactors must be carried out in order to more precisely define the viability of this technology.

### 6.2.7. *Feasibility of intensifying the production of vinyl acetate monomer*

Today, ethylene is the top choice for the manufacturing of vinyl acetate.

#### 6.2.7.1. *Conventional process*

Vinyl acetate is produced from ethylene, acetic acid and oxygen through the acetoxylation of ethylene in gaseous phase and catalyzed by palladium. The synthesis section is made up of a fixed bed multitube reactor (tubes with a diameter of 25 mm), operating at a pressure of 0.5–1.2 MPa and a reaction temperature of the order of 140–180°C. Ethylene is saturated in acetic acid and then introduced into the multitube reactor. The reaction is highly exothermic. The largest sub-product is carbon dioxide. The accumulation of secondary products is controlled by partial separation in order to maintain a constant rate. The vinyl acetate and acetic acid are separated by condensation/washing of the treatment flow. The remaining gaseous flow, largely made up of ethylene, is recycled, saturated with acetic acid in the evaporator and heated to the reaction temperature. The gaseous flow is then mixed with oxygen before being reintroduced into the reactor containing the catalyst. The gaseous reaction mix contains a 10–20% molar concentration of acetic acid, 10–30% of carbon dioxide (in moles) and around 50% of ethylene (in moles). The maximal oxygen content in moles is around 1.5% below the lower inflammability limit, which varies based on the composition of the gaseous mix and the reaction conditions.

#### 6.2.7.2. *Intensified process in a microstructured reactor*

Klais et al. (2010b) reported and commented on the content of an international patent filed by the company Udhe GmbH, then a subsidiary of the Hoechst chemical group. The patent concerned the manufacture, in gaseous phase, of vinyl acetate monomer improving the productivity of the process without any increase in the secondary products. The oxygen content in the reaction mix can be increased up to the explosive range without any risk of breaching the integrity of the reactor. There are, of course, no details given on the reactor, catalyst or the safety concerns involved. However, the process does contain many hidden risks, such as operating in the explosive domain, the composition of the primary materials and the condensable products, the dimerization of acetic acid, etc.

For the purposes of what follows, it is assumed that conventional equipment carries out the mixture and the saturation of the flow of reactive gas with ethylene and acetic acid (sections without the explosive melange). A microdesign mixture is then used to mix the gas flow saturated with oxygen. Finally, the reactor is made up of a microdesigned apparatus with multiple channels. Figure 6.8 presents the scheme for the principle underlying the intensified manufacturing process for vinyl acetate monomer. The mixing section thus has three compartments:

– The conventional mixer, M1, for mixing the recycled gas flow with fresh ethylene.

– The conventional mixer, M2, for saturating the gas flow from M1 with acetic acid vapor. The risk analysis assumes that the molar content of acetic acid in the gas flow is the same at equilibrium conditions at the given pressure and temperature. Any change in pressure or temperature can significantly modify the composition of the reactive gaseous mixture.

– The microstructured mixer, M3, makes it possible to obtain the required oxygen content in the reactive gas flow.

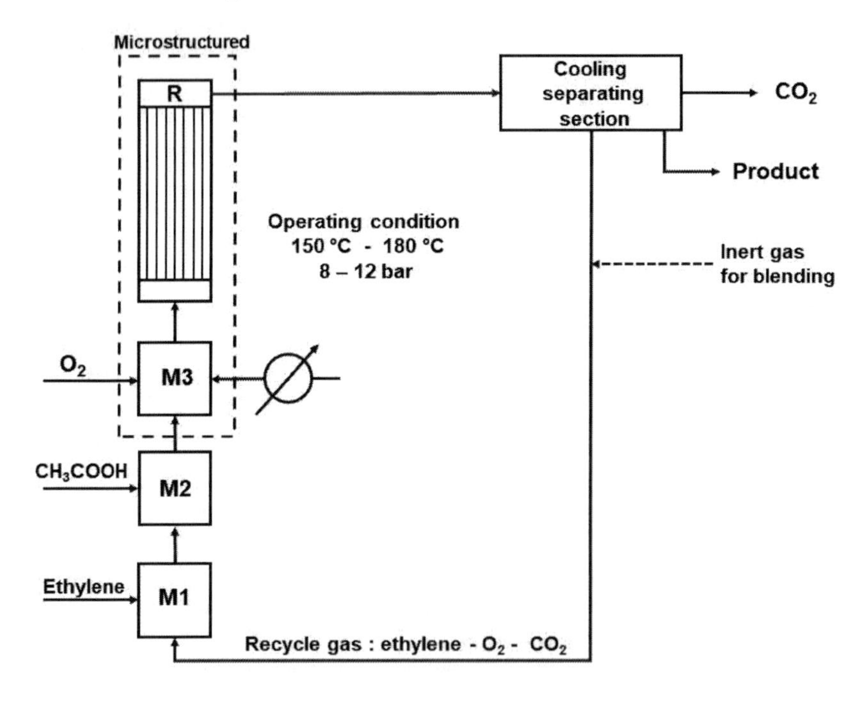


**Figure 6.8.** *Scheme showing the principle behind the intensified process for manufacturing vinyl acetate monomer (Klais et al. 2010b)*

The microstructured reactor, R, may contain a single microreactor or a network, in parallel, of microreactors with one or more microchannels 3 mm in diameter. The solid catalyst is deposited on the internal surface of each microchannnel.

## 6.2.8. *The microstructured reactor with catalytic walls: accelerator of the performance of a conventional tubular reactor*

In the examples presented in sections 6.2.4.2, 6.2.5 and 6.2.7.2, the innovation of the intensification processes was carried out directly in a microstructured reactor by partially or completely replacing an existing industrial reactor in a manufacturing unit. This new example aims to demonstrate the role of a microstructured reactor in association with the operation of an existing industrial unit. Let us look at the manufacture of phtalic anhydride through the controlled oxidation of orthoxylene in gaseous phase.

### 6.2.8.1. *Conventional process at moderate temperature*

The synthesis of pthalic anhydride is carried out in the presence of a catalyst made of vanadium pentoxide doped with titanium dioxide (Lepeu 2001). Air and orthoxylene are introduced, at a temperature of 423 K, in gaseous form at the head of a vertical fixed bed tubular reactor at an air/orthoxylene mass ratio of the order of 20. Each tube, having a diameter of 21 mm and height of 2 m, contains catalyst beads that are between 4 and 7 mm in diameter. The reaction is very exothermic, with a reaction enthalpy of 15.0 MJ per kilogram of orthoxylene. The coupling temperature of the reaction, said to be a moderate temperature reaction, is 640 K. The hot points in the reactor can reach 700–720 K in the upper quarter of the catalytic mass. The bundle of tubes is immersed in a bath of molten salts in a temperature range of 600–670 K in order to remove the heat produced by the reaction.

Becht et al. (2009) considered the advantage of highlighting three characteristic elements in the process that could make it possible to potentially reconfigure the process using a microstructured reactor:

– the distinct and highly exothermic nature of the process leads to significant limits on eliminating reaction heat and leads to pronounced temperature gradients throughout the reactor;

– the selectivity of the process, which is around 80%, can be improved. If you consider that the manufacturing cost of the xylene load is much greater than 70%, this could have a considerable economic impact;

– the xylene/air ratio defines the efficiency of the process and is limited by safety issues related to the risk of explosion. Although modern processes already function beyond inflammability limits, with typical xylene loads that can go up to 100 g/m$^3$ STP, this range could be further extended by using microstructured reactors with enhanced safety standards.

#### 6.2.8.2. *Reconfigured, intensified process*

Becht et al. (2007, 2009) considered that using a reactor with microstructured, catalytic walls offered an interesting option for modernizing existing installations. The guiding principle of this "booster" concept lies in the combination of a reactor with microstructured catalytic walls placed upstream of an existing conventional multitube reactor, as depicted schematically in Figure 6.9(a).

The heat transfer capacities and performances of the microstructured reactor must allow, among other things, the capping of heat points in the upper part of the conventional fixed bed tubular reactor. Figure 6.9(b) makes it possible to illustrate, through modeling, the potential contribution of the booster concept. In this example, the effective volume of the microstructured reactor corresponds to 20% of the effective volume of the fixed bed tubular reactor, where effective volume is defined as the product of the reactor volume and its spatiotemporal yield. Studying Figure 6.9(b) makes it possible to observe the following behaviors:

– Curve A represents the production rate for a multitubular reactor, with the correlated quantity of heat released, $\Delta H$. The production rate shows a steep increase in the first quarter of the reactor, with a corresponding production of heat. As the heat transport is limited in conventional multitube reactors, the production of heat limits the possible conversion level.

– Curve B corresponds to the evolution of the production rate of the microstructured reactor with catalytic walls in the case where the heat transfer is not limited to the interior of the walled reactor. The maximum rate of production and, consequently, the maximum heat released, $\Delta H_1$, will make it possible to substitute this for the upper zone with hot points from the conventional reactor.

– Curve C shows the full potential of the concept of the booster. A 250% increase in conversion could be produced, if the same amount of heat, $\Delta H$, was released in the multitube reactor, while adding only 20% of an increase in the nominal capacity because of a reactor with microstructured walls.

According to the numerical simulation, it would then be theoretically possible to make the multitube reactor operate at its maximum capacity, in order to increase the production rate, by either using more active catalysts or by increasing the temperature, pressure or gas flow, or by overcoming the presence of inert gases. Becht et al. (2009) carried out an estimation of the economical potential savings of this kind of intensified reconfiguration using a booster connecting a microstructured reactor and a conventional reactor.

### 6.2.9. *Generic example of direct gaseous fluorination of a liquid hydrocarbon*

In a prospective review of the "molecular processes-product-process" triplet in connection to the future of process engineering, Charpentier (2002) recommended the use of structured microreactors to open new reaction pathways that were considered too difficult to implement in conventional industrial equipment. In terms of intensification, he suggested exploring the direct fluorination of aromatic compounds.

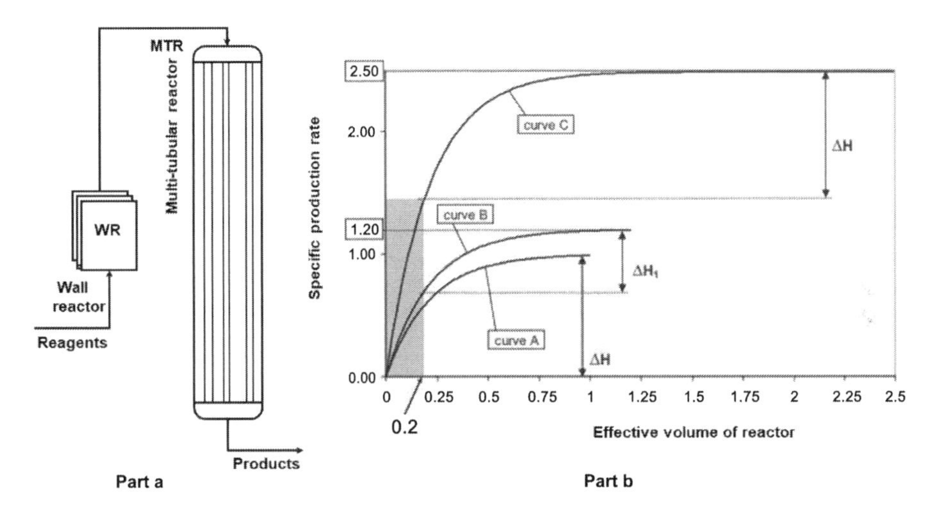


**Figure 6.9.** *(a) Arrangement of a microstructured reactor with catalytic walls and a conventional tubular reactor; (b) comparisons of their performance (Becht et al. 2007)*

Lantz (1998) described the different industrial fluorination processes using hydrofluoric acid, elementary fluorine or metallic or non-metallic fluorides. Fluorinated aromatic compounds are of great commercial interest, for instance, pharmaceutical products, reactive dyes, isotopic tracers and liquid crystals. Two pathways are prioritized for synthesizing these products industrially: the Schiemann process and the Halex process, both of which use fluorinating agents. The yield from both processes is limited as the synthesis proceeds in several consecutive steps (Schiemann) and as there are secondary reactions due to the long reaction time (Halex). A single-step process, like direct fluorination using elementary fluorine, would potentially offer a number of advantages: greater yield, easier treatment and greater flexibility in the process for the different reactants, less energy, less waste, simpler equipment and cheaper primary materials (Schuster 2005).

However, the characteristics of the process still remain problematic, as can be seen in the following list:

- the reactants are dangerous;
- the system is multiphase (gas–liquid reaction);
- the reactions are highly exothermic (rapid radical reactions);
- the formation of multiple products (mechanism of the series-parallel reaction);
- low conversion due to reactions inhibited by mass transfer.

The key points to ensure safe conditions, higher yield and higher conversion reside in the integrated elimination of heat and effective and controlled gas–liquid contact. Many techniques have been proposed to enhance control over the reaction:

- diluting the fluorine using an inert gas (nitrogen, etc.);
- diluting the organic product to be fluorinated using an inert solvent;
- fluorination at low temperatures;
- continuous processes with concentration and temperature gradients.

Chen et al. (2008) described four different types of gas–liquid microreactors: microchannel contactors, falling film microreactors, microreactors with phase interface stabilized by physical structure and packed bed microreactors. The hydrodynamic behaviors and heat and mass transfer capacities of each were specified. For the generic example proposed here, it is interesting to study the work published by Jahnisch et al. (2000) and Schuster et al. (2008).

Jahnisch et al. (2000) studied the functioning of a falling film microreactor, 10 $\mu$m in thickness, generated on a grooved, vertical plate containing a large number of parallel channels. Each channel has a cross-section of 100 $\mu$m x 200 $\mu$m. The microreactor is fed by fluorine, in gaseous phase, either pure or diluted in nitrogen. The liquid phase is made up of toluene or a mixture of toluene and solvents such as acetonitrile or methanol. The feasibility of direct fluorination has been discussed with the study of experimental selectivity-conversion diagrams, the composition of the products and derivatives formed, the influence of the fluorine–toluene ratio, the temperature, the nature of the solvent and the residence time of the liquid phase. The results that were obtained showed the beneficial effect of the improved heat and mass transfer in the microreactor.

Schuster (2005) and Schuster et al. (2008) considered the respective operations of a falling film gas–liquid microreactor and a gas–liquid microreactor with a catalytic membrane Figure 6.10 depicts a cross-section of the two gas–liquid microreactors studied.

The system studied here is also the direct fluorination of liquid toluene. The falling liquid film circulates by gravity in the microchannels. With respect to the liquid flow, the gaseous phase can circulate concurrently toward the base or countercurrent upwards. The direction of circulation of the coolant, with respect to the reactant liquid flow, can be either concurrent, countercurrent or cross-current. The authors developed a model of a falling film microreactor through a mathematical description of the equations for conservation of heat, mass and momentum in 3D modeling. The model for a microreactor with passive or catalytic membrane can be considered to be an extension of this aforementioned model. The 3D model was validated with experimental results from Jahnisch et al. (2000). The effects of heat transfer and mass transfer in these microreactors could thus be numerically described and compared to reactors in a pilot phase and conventional industrial reactors. For example, for a 500 $\mu$m x 100 $\mu$m rectangular microchannel, in terms of mass transfer, the coefficient for the mass transfer, on the liquid side, and the interfacial area per unit volume of the microreactor can reach 21 $s^{-1}$ and 9,000 $m^2.m^{-3}$, respectively. These values clearly exceed those of conventional contactors such as bubble column reactors or packed column reactors. In terms of heat transfer, the heat transfer coefficient between the flowing liquid and the wall of the channel where it is flowing is of the order of 2,800 $J.m^{-2}.s^{-1}.K^{-1}$.

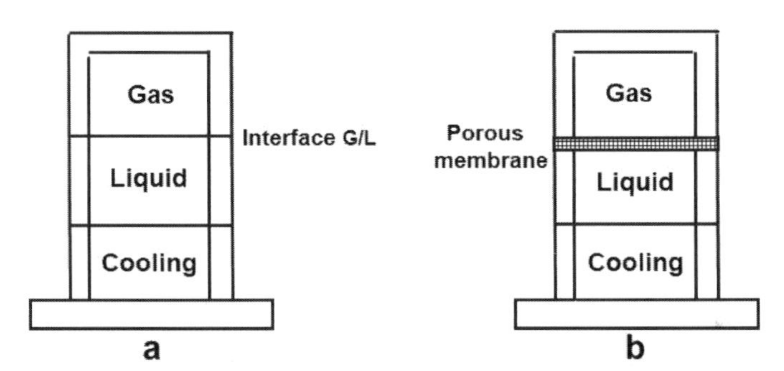


**Figure 6.10.** *Sketch of the configurations of a falling film microreactor (a) and with a passive or catalytic membrane (b) (Schuster et al. 2008)*

To conclude this section, looking at examples of process intensification, there are other additional examples of intensification in a broader sense, listed in the excellent review by Hessel et al. (2013). These authors examined different paths for new perspectives of innovative processes using operating conditions that were removed from conventional practices. The ways described concerned the use of high temperatures (Hessel et al. 2011), high pressures (Hessel et al. 2011), strong concentrations (without solvent), new chemical transformations (Hessel 2009), explosive conditions, as well as the simplification and integration of processes to stimulate chemical synthesis, both at the laboratory level as well as on an industrial scale. These methods are grouped in different categories based on the chemical processes and their designs. The specific use of structured microreactors to safely implement difficult and dangerous designs is highlighted, because of the excellent properties of the intensification of transfers.

| **Elementary limitations** | **Complex limitations** |
|---|---|
| Fluid-solid mass transfer | Fluid or equipment volume |
| Fluid-fluid mass transfer | Non-uniform properties or conditions |
| Mixing single-phase mass transfer | Size distribution |
| Heat transfer | Difficult activation |
| Kinetics | Saturation effect |
| Thermodynamic phenomena | Safety |
| Residence time distribution | Pressure drop/mechanical energy |
| Dynamics, inertia, transient effects | Energy consumption<br>Utility consumption |

**Table 6.2.** *Inventory of simple and complex limitations (Commenge and Falk 2014)*

## 6.3. An attempt to rationalize intensification equipment

The earlier review of certain intensification processes, with respect to their "inherent" safety, has shown the diversity, variety and even the abundance of innovative solutions being proposed. However, it would be advisable to propose a decision-making methodology to help quickly select the best of available technology, from among the various potential solutions, in order to choose and try to rationalize the emergence of proposals.

Commenge and Falk (2014) thus proposed a methodological screening framework for choosing intensified equipment and for the development of innovative technology. The methodology is based on the concept of analysis of characteristic times for the steps in a process. This analysis makes it possible to establish a relationship between the characteristic dimensions and the efficiency of the physical and chemical processes involved. It also gives room to demonstrate how the concerned reactors may act upon the hierarchy of phenomena that control the processes governing the system's efficiency. The schematic structure for this method is depicted in Figure 6.11.

Once the initial problem to be resolved is defined (synthesis, reactions, process), the first step consists of listing the various limitations of the process. These are divided into elementary and complex limitations, covering a wide spectrum of cases (Table 6.2).

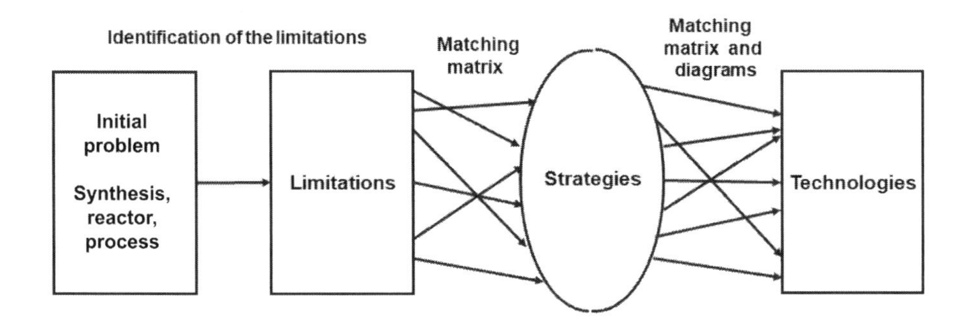


**Figure 6.11.** *Schematic structure for the methodological selection method*

The details of the different limitations are described and characterized. Various figures are given as examples to illustrate the results of the procedure:

– the comparison of different equipments based on the characteristic gas–liquid mass transfer time;

– the comparison of the mixing times of different reactors;

– the comparison of the characteristic reaction times of the various non-catalytic organic syntheses in liquid phase in the microstructured reactors.

The complex "safety" limitation is understood by examining the following factors (Carvalho et al. 2008):

– reaction enthalpies: limitation studied through heat transfer, inflammability, explosiveness (geometric microstructuring would enable functioning in explosive regime conditions), modifiable operating conditions such as temperature, pressure and composition;

– toxicity: the modification of phases or properties of fluids may be considered, as well as geometric microstructuring to enhance the stability of emulsions without additives;

– corrosion: the modification of phases or properties of fluids, the modification of materials and operating conditions may be applied;

– inventory of fluids: examining the limitation with respect to the fluid volume/equipment volume ratio;

– temperature: operating conditions may be modified, but the limitation must also be studied in terms of heat transfer;

– pressure: new operating conditions and geometric structuring could be solutions;

– process structure and configuration: there are many possible modifications to the structure and configuration of a process. For example, multifunctional reactors and geometric structuring, with special attention to interactions between the different constituent units of the process.

It must be noted that Carvalho et al. (2008) recommend the use of performance indicators to assess the complex "safety" limitation.

The schematic representation in Figure 6.11 indicates the presence of a block of strategies. Commenge and Falk (2014) have thus simultaneously proposed a classification of these strategies, as presented in Table 6.3.

The second step in the methodology uses a pre-filled connection matrix to associate the identified limits with a set of intensification strategies, like geometric microstructuring, period operation or multi-scale design. In the matrix, the elementary and complex limitations are presented in rows, while the strategies correspond to the columns. Each cell is attributed a score, ranging from 0 (no impact) to 5 (heavy impact). This describes the pertinence of that strategy with respect to the limitation.

| Molecular scale | Equipment scale | Process scale |
|---|---|---|
| Operating conditions (pressure, temperature, concentration) | Geometric structuring | Parallelism, multi-scale |
| Change of phases/solvents | Catalyst structuring | Segmentation |
| Fluid phase properties | Gravity, centrifugal forces | Periodic operation |
| Inert species addition | Shear rate | Alternative energy sources |
| – | Material properties | Coupling with separation |

**Table 6.3.** *Inventory of intensification strategies (Commenge and Falk 2014)*

The third step connects these strategies to a list of technologies that apply these strategies or within which they can be applied. Matrices make it possible to sort these technologies by relevance to the initial problem. The final step gives quantitative diagrams to compare the characteristics of these potential solutions with specifications of the problem. The methodology not only makes it possible to obtain a restricted list of appropriate solutions to be considered in terms of technical design and economic assessment, but also provides a list of solutions to implement.

The application of this methodology is illustrated and discussed in the example of the manufacture of a fine chemical organic compound through the organometallic Grignard reaction. The very exothermic reaction presents secondary products generated by parallel and consecutive reactions.

## 6.4. Concept and application of a general methodological framework for the synthesis and design of processes that integrate intensification

Lutze et al. (2010, 2013) studied the similarity of the molecular and process scales and considered that the intensification of a process could be envisaged at the level of the process as a whole, at the scale of the different unit operations, or elementary functions and/or phenomena, as depicted in Table 6.4.

| Scale | Parallel of similarities | | |
|---|---|---|---|
| **Molecule** | Molecules | Functional chemical groups | Atoms |
| **Process** | Industrial unit | Unit operations | Basic phenomena |

**Table 6.4.** *Simplified diagram of the similarities on the molecular and process scales (Lutze et al. 2013)*

By prioritizing an approach at the phenomenon scale, a new methodology makes it possible to generate options for potentially novel, intensified processes. The authors thus proposed the concept for a general, systematic framework containing hierarchic steps to analyze an existing process, in order to create and assess these options. Each step requires the use of tools and predictive methods. The solutions obtained are simultaneously valued using environmental, economic, safety and intensification metrics. Lutze et al. (2010) applied the method to the case study of the catalytic synthesis in liquid phase of isopropyl acetate, an important industrial solvent. The balanced, equimolar reaction, in liquid phase, of isopropanol and acetic acid is translated by the global equation:

$$CH_3COOH + C_3H_7OH \longleftrightarrow C_5H_{10}O_2 + H_2O$$

The existing process, considered to be the standard, is carried out in a batch reactor. The two loads of isopropanol and acetic acid placed in the reactor react at a temperature of 343 K for a duration of 1,500 min. The reaction mixture obtained is then distilled to eliminate acetic acid. The distillate contains a ternary azeotropic mixture of isopropyl acetate, isopropanol and water. The isopropyl acetate is finally purified at a content of at least 90% by weight through extractive distillation with the appropriate solvent.

The two final options chosen to improve the process by applying the proposed procedure are related to the coupling of the batch reactor, either with an external pervaporation unit operation with recycling, or with a pervaporation unit operation in situ, in the batch reactor. Table 6.5 presents the comparison of the three processes for the same chemical synthesis using different performance metrics (Lutze et al. 2010).

When respecting the enforced parameters for reaction and pervaporation temperatures, the catalyst concentration, the isopropyl alcohol/acetate acid ratio, and the fixed area of the membrane, the simulation and optimization results given in Table 6.5 show that the objective functions of the configurations with the membrane are the best options. Indeed, the initial objective function of the standard for the batch reactor changes from 2.7 g of reactant per gram of product to a value of 1.51 g/g for options with a membrane. Similarly, the yield from the synthesis reaction changes from 74.4% to 90% with the same product quality. The other performance metrics are also affected by the addition of the membrane separation unit operation. For example, the operating costs are reduced because of the reduced use of reactants, catalyst and heating energy.

Lutze et al. (2013) complemented their work with a study of the same synthesis of isopropyl acetate using the same method in other conditions. The basic preliminary draft was that of a perfectly continuously stirred reactor that would produce 50,000 tons per year of isopropanol acetate. The reactor operates at an isothermal temperature of 300 K under a pressure of 1 bar. The continuous feed is an equimolar mixture of two reactants, whose respective, identical streams are 1,430 moles per minute. Since the balanced reaction is incomplete, the residual output streams of each reactant are 499 moles per minute. The exothermic reaction requires a total removal of the heat of 16.7 MJ per minute. The authors examined 22 different configurations of the reactors. Through the minimization of the objective function, they demonstrated that to theoretically obtain a conversion rate of 90% with a purity of 98% isopropyl acetate, the best option was an integrated unit operation comprising a fixed-bed continuous open plate heat exchanger-pervaporator catalytic reactor and frames with rectangular cross-section. The enforced initial production would require the arrangement in parallel of 20 integrated devices.

Lutze et al. (2013) compared the simulated results obtained with this new design and the data from literature obtained for a reactive distillation system (Lai et al. 2007) and for a basic conventional process (Corrigan and Stichweh 1968). The reactive distillation system comprises a reactive distillation column, equipped with a decanter at the head of the column and an external stripping column. The distillate from the reactive distillation column is close to the composition of the ternary azeotrope. The final product is obtained as the bottom product of the stripping column. The conventional process chosen is more complex with regard to the number of unit operations, as the reaction is incomplete and separation is difficult because of the number of azeotropes. This process consists of a reactor, six distillation systems, an extractor using water as solvent and a decanter. The authors report that the unit, dimensioned using the synthesis/design methodology based on phenomena would lead to a conversation rate of 99% of isopropanol acetate, while the respective values would be 93% and 94% for the conventional process and for the reactive distillation unit. The thermal energy requirement for the new design would only be one-tenth of the requirement of the conventional process and around one-fifth of that of the reactive distillation process. Finally, the amount of catalyst in the new design could be reduced to 40% with respect to the design of the reactive distillation system.

| Items | Parameter | Batch reactor | Internal reactor pervaporation | External reactor pervaporation |
|---|---|---|---|---|
| Process conditions | Reactor temperature (K) | 343 | 343 | 343 |
| | Pervaporation temperature (K) | – | 343 | 343 |
| | [catalyst] % weight | 10 | 10 | 10 |
| | [$C_3H_7OH$] (mole) | 3.66 | 1.9 | 1.9 |
| | [$C_3H_7OH$]/ [$CH_3COOH$] | 1.3 | 1.3 | 1.3 |
| Objective function | g reactant / g product | 2.7 | 1.51 | 1.51 |
| Economic metrics | Operation cost | 0 | lower quantity of materials, catalyst, lower heating | *Idem* |
| | Capital cost | 0 | smaller apparatuses, fewer process steps, but increased pervaporation membrane cost | *Idem* |
| Safety metrics | Safety | 0 | Smaller volume, smaller streams | *Idem* |
| Environmental metrics | Energy | 0 | Less energy for heating, no energy intensive distillation column | *Idem* |
| | Yield from reaction (%) | 74.4 | 90 | 90 |
| Intensification metrics | Volume | 500 $cm^3$ | Smaller reactor, single step process | Smaller reactor and smaller pervaporation units |
| | Simplification of process | 3 units | 1 unit | 2 units |

**Table 6.5.** *Comparison of the three units for the synthesis of isopropyl acetate based on performance metrics (Lutze et al. 2010)*

## 6.5. Reality or myth? Safety 4.0 in intensification processes

The observation of the wave of innovation for intensification processes has led to certain qualifiers with regard to their safety, for instance, safer processes, eigen safety processes, inherently safe processes, intrinsically safe processes, intrinsically safer processes, etc. These different terms may cover a certain number of novel concepts related to the integration a priori of safety into the design and operation of industrial processes (Amyotte et al. 2009). What do these qualifiers mean? For example, the dictionary definition of "intrinsic" is belonging to an extremely important and basic characteristic of a person or thing, independent of external factors. Safety will thus be included in the process itself from the basic design onwards, rather than being added as an afterthought by adding active, passive or procedural measures. How can the intrinsically safer characteristic of intensification processes be evaluated? Are ISD principles effectively applied here? There are many complicated responses to discuss to gauge the efficiency and added value a priori of ISDs in intensified processes.

### 6.5.1. *A few assessment tools*

There are several assessment tools and methods in literature, a review of which is given in Khan and Amyotte (2003).

#### 6.5.1.1. *Performance indicators*

Section 5.15.3 in Chapter 5 described a few recommended examples of performance indicators for process safety. Let us recall that the commonly used hazard metrics or indices are as follows: the Dow fire and explore index, the Dow chemical exposure index, the Mond index of fire, explosion and toxicity, the weighted hazard index for fire, explosion and toxic waste, the toxicity index, the assessment tool for managing environmental hazards (air, surface water, underground water, used water), the risk assessment model related to the transport of chemical products and the hazardous waste index (Khan and Amyotte 2003).

The school of thought promoted by A.M. Heikkilä of the Helsinki University of Technology (Finland) proposed a new index for inherent safety to assess the global contribution of the inherent safety of a process (Heikkila et al. 1996; Heikkila 1999). This relatively simple index is designed to take into account a series of factors that affect global inherent safety. These factors are grouped into two main categories: chemical safety and process safety. The index for inherent chemical safety describes the effect of the choice of primary materials and other chemical products, taking into account the following elements: reaction enthalpies, inflammability, explosivity, toxicity, corrosivity and the incompatibility of chemical products. The index of the inherent safety of the process describes the impact of the type of equipment and process conditions on safety. The parameters taken into consideration are the list of chemical products, temperature and pressure, the type of processing equipment and the process structure. The separate indices thus obtained are added together to calculate an estimation of the total inherent safety index.

Sultana and Haugen (2022) recently presented an inherent safety index for a system to evaluate an inherently safer design from the phase that is upstream of the development of process. The analysis begins by identifying the characteristics of the inherent safety of an ideal, non-hazardous system, along with the associated parameters. Four sub-indices are established, determined from the characteristics of the non-hazardous system by using their relevant parameters. The safety of the chemical processing system, worker health and environmental safety can be ensured by selecting the relevant parameters. Table 6.6 provides an overview of the characteristics, conditions and parameters of an inherently safer ideal system.

| **Characteristics of the inherently safer system** | **Conditions related to the inherent safety** | **Inherent safety parameters** |
|---|---|---|
| Safe inflow to the system | Safer material inflow | Chemical, physical and external properties of the material (flammability, chemical instability, corrosivity, viscosity, phase, quantity, or mass) |
| | Less energy consumption by the process and equipment | Energy consumption by the process<br>Energy consumption of the equipment |
| | Higher efficiency of the processes or equipment | Efficiency of equipment |
| Safe production of the system | Safer intermediate product or by-product | Chemical, physical and external properties of the by-product and intermediate products |
| | Safer energy production | Heat of reaction |
| | Less production of waste material | Amount of waste material |
| | Less production of emission | Amount of greenhouse gas emission<br>Amount in the form of $CO$, $CO_2$, $SO_2$ etc. |
| Simple | Simpler processes and individual components and procedures | Process complexity parameters |
| Non-vulnerable | Safer process | Presence of unique hazardous process |
| | Compatible | Hazardous interaction between various parameters |
| | Safer process condition | Extreme hazardous condition |

**Table 6.6.** *Summary of the characteristics, conditions and parameters of an inherently safer ideal system (Sultana and Haugen 2022)*

The parameters are quoted based on their deviation with respect to the ideal safety condition. The sum of the deviations of the parameters gives the value of the inherent safety index. A case study applied to the evaluation of six possible chemical pathways in the production process of methyl methacrylate is given. The details of the computation of the inherent safety indices are given in the additional data provided in Sultana and Haugen (2022).

#### 6.5.1.2. *Methods dedicated to the inherent safety of a process*

Several conventional risk assessment methods for an industrial process are presented in Table 3.3 in Chapter 3 and in Chapter 4. The existence of principles to control risks, including the intensification principles, offering pathways to develop inherently safer processes, involved re-assessing, adapting and modifying traditional risk assessment methods in order to ensure they remain relevant and effective. To this end, two important multi-year industrial research programs have been initiated and supported by the European community.

##### 6.5.1.2.1. The INSET tool from the INSIDE program

The three-year (1994–1997) research program INSIDE (INherent She In Design) was initiated and supported by the European Commission (DG XII). This project aimed to promote the adoption of inherently safer approaches in designing and developing industrial chemical processes. This research process was led by a joint team drawn from the industry and research bodies made up of AEA Technology, TNO, VTT Manufacturing Technology, Eutech Engineering Solutions, INBUREX, Kemira Agro and ICI Polyurethanes. The results from the INSIDE project were published in the form of an integrated two-volume (481 pages) collection of tools and methods, titled *INherent She Evaluation Tool* (INSET) (INSET 2001). A brief summary of the contents is given as follows:

– detailed constraints and objectives analysis;

– process option generation (including process waste minimization guide);

– preliminary chemistry route options record;

– preliminary chemistry route rapid ISHE evaluation method;

– preliminary chemistry route detailed ISHE evaluation method;

– chemistry route block diagram record;

– chemical hazards classification method;

– record of foreseeable hazards;

– performance index for Safety, Health and Environment (ISHE):

  - fire and explosion hazards index,

  - acute toxic hazards index,

  - health hazards index,

- acute environmental incident index,
- transport hazards index,
- gaseous emissions index,
- aqueous emissions index,
- solid wastes index,
- energy consumption index,
- reaction hazards index,
- process complexity index,

– multi-attribute ISHE comparative evaluation;

– rapid ISHE screening method;

– chemical reaction reactivity – stability evaluation;

– process SHE analysis/process hazards analysis and ranking;

– equipment inventory functional analysis method;

– equipment simplification guide;

– hazards range assessment for gaseous releases;

– siting and plant layout assessment;

– designing for operation.

This framework provides chemists, process engineers and safety managers with the tools required to identify, evaluate, optimize and systematically select the designs and inherently safer HSE processes. This may be applied to projects with an entirely new process, to an existing process in a new factory, or to modifications made to an existing factory or process. It must be highlighted that the procedure examines risks related to safety, health and the environment in an integrated manner in order to guarantee that the conflicts and synergies between these aspects are recognized and efficiently managed.

The proposed procedure is versatile and flexible. It focuses on the first key steps in a project where almost all major decisions are taken and which determine the HSE performances of the factory:

– Step I : selection of the chemical pathway;

– Step II: detailed evaluation of the chemical pathway;

– Step III: optimization of the process design;

– Step IV: factory design.

Detailed examples of how this can be applied to the choice of a preliminary chemical pathway and to the practical assessment of the selected pathway in the field of fine chemistry are given, as well as examples of application to continuous production in order to increase the production capacity of an intermediate product of the ICI company.

#### 6.5.1.2.2. The deliverable "tools" of the IMPULSE project

We have already seen the European project IMPULSE in section 6.2.4. To recap, in the framework of this project, a working sub-group was given the specific mission of developing new methods and tools to analyze, evaluate and control accidental risks induced by the intensification of chemical processes. The contents of the deliverable *Guidance on Safety/Health for Process Intensification including MS Design* were published as four articles in a special edition of the journal *Chemical Engineering and Technology* (Klais et al. 2009a, 2009b, 2010a, 2010b).

In the first article, Klais et al. (2009a) listed the hazards related to the reactions that were carried out in the intensified processes. The use of process intensification in multi-scale equipment has a deep impact on the manner in which chemical products are produced. The shift to higher spatiotemporal yields, at higher temperatures and confined reaction volume involves new risks, such as thermal runaway, consecutive decomposition, the formation of hot spots and the incomplete conversion of reactants. Using simplified spreadsheet calculations, the authors estimated the temperature profiles, conversion rates and consequences of any potential dysfunction in the kinetics of the reaction. The results from this analysis show that the range of optimal reaction conditions is almost congruent with the danger of a non-controlled reaction. It may be assumed there is a risk of a spontaneous reaction with hot spots if highly exothermic reactions are carried out in micro-designed reactors.

In the worst case, thermal runaway is followed by decomposition, with the release of non-condensable gases. The estimations show that a microreactor is not likely to be at risk of overpressure as long as at least one end of the reactor is not blocked. Nonetheless, the design concept for inherently safe reactors using microreactors is not valid in general (de Graaf and Tikku 2007). Indeed, the ISD concept can only be satisfied for reactions with a moderate reaction heat, where the reaction velocity constant is low at the operating temperature, and where the residence time is adapted to the microreactor under consideration. In these conditions, the system's thermal inertia may constitute a thermal sink and reduce the potential auto-acceleration of the reaction.

Klais et al. (2009b) studied the difficult problem of assessing the risk of explosion in the microstructured equipment. The question of explosion is discussed here in terms of the inherent safety of microdesigned equipment, either to put out inflammation, or to suppress the propagation of the explosion. The presence of mixtures of inflammable gases or the use of solid catalysts deposited on surfaces are examples of potential permanent sources of inflammation, even at temperatures below the auto-inflammation temperature. When discussing the data published in open literature, and especially comparing flame quenching distances with the characteristic dimensions of microstructured channels, Klais et al. (2009b) concluded that microreactors are inherently safe with regard to the extinguishing of the ignition process within the microchannel, even in the presence of a catalyst or hot spot. However, to be vigilant, it must be considered that multi-scale equipment is not systematically inherently safe. Indeed, the reverse must be considered, where the propagation of an external explosion can enter the same microreactor. In this case, the external explosion may destroy the microreactor, while for the same initial conditions, an explosion may not be generated within the microreactor. This apparent paradox in how the microreactor behaves with an internal or external explosion can be explained as follows:

– the microreactor acts as a resistance to the flow and, consequently, the adiabatic compression of the non-combusted gas upstream of the flame front is possible, as well as the stacking up of overlapping pressure waves;

– in narrow channels, the turbulences created by high flow velocities could cause premature ignition, resulting in the explosion accelerating in a manner comparable to detonation.

The content of the risk assessment published by Klais et al. (2010b) focuses on controlling potentially hazardous situations directly related to the new, multi-scale process intensification technologies. The authors first agreed on identifying potentially hazardous situations that could produce uncontrolled reaction conditions or explosions. Next, the potential causes for these deviations were discussed, in relation to the negative impacts on health, environment and the process. An important aspect of process safety involving new technology is the interface with traditional equipment, with respect to greater section/volume ratios and command devices that are ineffective with the new technology. This interface essentially determines the performance of the intensified equipment. Deviations in the operation of the adjacent equipment, upstream and downstream of the intensified equipment, are also important for the safe functioning of the whole process. Consequently, the integration of new technology in a conventional production unit must be carefully analyzed for every failure of the interface between the conventional equipment and the micro-designed equipment. In order to take into account the characteristics of multistructured equipment, Klais et al. (2010b) developed a risk analysis method, called the HAZOP-LIKE method, which created a synergy between the HAZOP method and the preliminary risk assessment method (PRA). According to the authors, the HAZOP-LIKE method has the following advantages:

– it is applicable to thermodynamic systems and to fluid flows;

– the basic concept of the HAZOP method is widely used in the industry;

– it is well structured and its systematic approach translates into a specific list of keywords to define the deviations to be studied;

– it makes it possible to categorize the priorities to improve plant safety;

– it can also be implemented from the earliest design stages, in order to assess potential disruption points as early as possible, with the goal of integrating the necessary safety measures;

– the method can also be used in later stages of the design process to verify that the risks are contained to an acceptable level.

Since the standard list of keywords, listed in the IEC 61882 (2016) for the conventional HAZOP method, is not adequate to identify specific safety risks for microstructured elements, the authors have developed a more appropriate list of parameters and keywords, the combination of which can make it possible to identify 21 possible deviations. These are given in Table 6.7. It must be noted that this list is not exhaustive and that the specific characteristics of the new, micro-structured elements must be taken into account in terms of the new parameters and keywords, if required.

The estimations of gravity and the probability of the severest consequences are evaluated on a scale with four levels: zero, weak, moderate and strong. The practical use of the HAZOP-LIKE method is illustrated by two examples. The first concerns a homogeneous, liquid–liquid, catalytic, exothermic reaction carried out in a microstructured installation comprising a mixer, a reactor and a thermal exchanger. The second is the production of vinyl acetate monomer in a microstructured reactor, as described in section 6.2.7.2.

The final contribution from Klais et al. (2010a) is a summary of knowledge relative to risk assessment for the microstructured processes studied earlier. The authors first recall the specificity of the characteristics and behaviors of microstructured and multi-scale processes and equipment:

– short residence times, leading to poor functioning during fluctuations in flow rate and pressure;

– the chemical conversion being highly dependent on minor variations in temperature and flow rate;

– a high area/volume ratio and small diameters, sites of high influence with respect to fouling and deposits;

– structure integrity that must be monitored, due to the use of new assembly technologies;

– influence of corrosion/erosion on the performance of microdesigned equipment.

| Parameter | Key words | Multi-scale (MS) specificities |
|---|---|---|
| Pressure | too high | issue of interface traditional/micro-equipment primarily |
| | fluctuation (over time) | on a time scale below seconds |
| | too low | avoid evaporation, flashing, two-phase flow |
| Temperature | too high | critical due to reaction conditions close to runaway/decomposition |
| Flow rate | too low | risk of incomplete reaction/condensation/crystallization |
| | too high, fluctuating/inhomogeneity (in time and space) | risk of incomplete reaction on the time-scale below seconds and on the space scale of the MS equipment |
| | too low/no flow adverse | partial or full blocking of single channels more probable |
| Concentration | wrong ratio of reactants fluctuation/inhomogeneity (in time and space) | reduced solvents more likely, on the time scale below seconds and space scale of MS equipment |
| Structural integrity | deterioration | scale of MS equipment |
| Secondary containment | loss of | alternative safety measure for MS equipment |
| Medium | other than correct (e.g. solid contaminants, impurities, detachment of catalyst, crystallization of reaction products) | respect scale of MS equipment |
| | leakage of cooling media into process stream | intrusion of heating/cooling medium |
| | leakage of process stream into cooling media | intrusion of heating/cooling medium |
| Reaction | too fast too slow other reaction products/side reactions | – |
| Cooling capacity | loss of or less | – |
| Utilities | loss of (e.g. electrical energy, cooling water, instrument air, inerting gas) | – |

**Table 6.7.** *List of guide or key words of the HAZOP-LIKE method for micro-equipment (Klais et al. 2010a)*

The study finally reports the risk assessment of two demonstrations in the IMPULSE project:

– the oxidation of dioxide by air in a block of microreactors with two reaction zones at different temperatures;

– the alkylation reaction to produce an ionic liquid in a multi-scale production unit, integrating a micro-mixer, a microstructured reactor and a ring-channel reactor.

The HAZOP-LIKE method was applied in both cases to assess potential risks, including those of inter-dependence, with respect to the conventional connected equipment.

### 6.5.2. *Examples of safety versus intensification conflicts*

Several studies (Hendershot 1995; Luyben and Hendershot 2004; and Etchells 2005) have highlighted the risk of considering a priori the label "inherently safe" for installations with processes arising from an ISD and especially for installations that implement intensification processes. It is essential to consider beforehand, as far upstream as possible, all industrial safety imperatives in order to ensure that the choice that is made does not lead to a process-intensification-safety conflict. Several examples illustrating this difficulty are described in this section.

#### 6.5.2.1. *Selecting a reaction solvent*

It is often necessary to choose from several solvents to carry out a reaction. The potential solvents present significant differences in terms of hazardous characteristics, such as toxicity, inflammability and volatility, for example. A good example is given in CCPS (2012). In the case of an exothermic reaction, it is possible to choose between a toxic, but non-volatile solvent, and a non-toxic but volatile solvent. Each solvent offers inherent advantages and drawbacks in terms of safety, as summarized in Table 6.8.

This seemingly simple example underlines the potential difficulty of this decision. There is no general answer to the question of whether we know which of these two proposed solvents is inherently safer. In fact, the advantages and drawbacks of each option must be compared for each individual case. The final choice must be made based on the specific details of the process, reactants, products and materials.

#### 6.5.2.2. *Classic situation of a continuous reactor compared to a semi-batch reactor*

The strategy of developing a continuous reactor is one that is often suggested to improve the safety of a process that is conventionally carried out in a batch or semi-batch reactor. Let us consider the following exothermic reaction, with instantaneous reaction kinetics in the presence of the catalyst C:

$$A + B \longrightarrow D + E$$

where the reactant A is in a solution in the solvent S. The higher risk with this process resides in the great instability of the reactive mass, if the product B accumulates in the reactor or if the catalyst C is missing by accident. If the concentration of reactant B exceeds a certain critical value, a potential explosion may quickly occur. Figure 6.12 presents the schematic principle for the process carried out in a semi-batch reactor (Figure 6.12(a)) and in an open reactor (Figure 6.12(b)).

| Solvent | Inherent safety advantages | Inherent safety drawbacks |
|---|---|---|
| Non-toxic and volatile | Solvent is non-toxic, reducing hazards in normal handling, and in the event of a discharge due to a runaway reaction; the volatile solvent limits the temperature rise in case of a runaway due to the tempering effect when the solvent boils | High vapor pressure of solvent results in potential for high pressure in the reactor in case of a runaway exothermic reaction |
| Toxic and non-volatile | Runaway exothermic reaction may not be sufficient to raise the reaction mixture temperature to its boiling point, so there is no hazard of over-pressurizing the reactor | Potential exposure of personnel to toxic solvent; environmental damage in case of a spill |

**Table 6.8.** *Some inherent safety advantages and drawbacks of alternative process solvents (Hendershot 1995)*

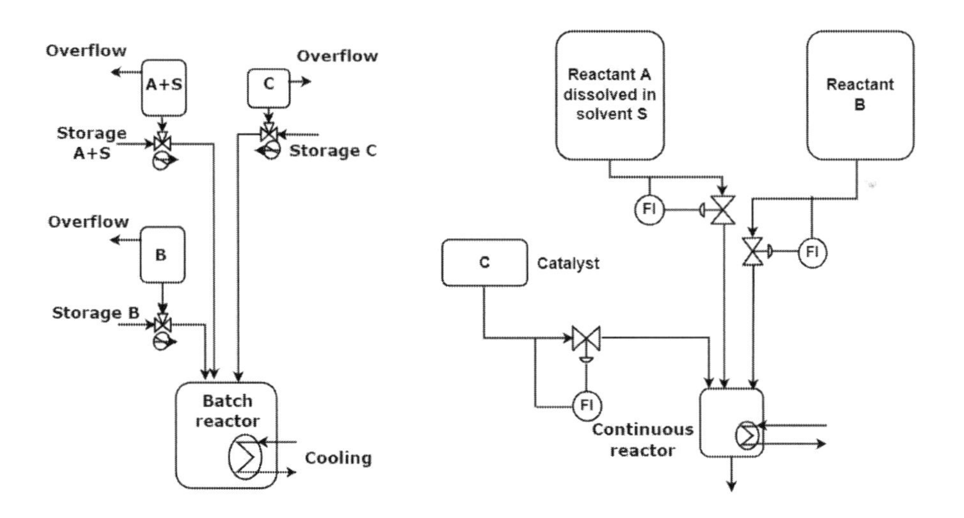


**Figure 6.12.** *Schema showing the principle of the semi-batch reactor (a) and the open reactor (b) (Hendershot 1995)*

In the discontinuous process, the correct quantity of reactant A in solution in solvent S is introduced into the reactor from a weighted reservoir dimensioned such that it contains the exact load for one batch. A three-way valve at the foot of the reservoir manages the direct flow from the storage tank into the reactor in bulk. An overflow makes it possible to send any possible excess of the A+S mixture introduced into each load. The catalyst C is then introduced. A release of heat due to the enthalpy of the mixture confirms that the load has indeed been added. The reactant B is then added at a controlled rate to maintain the desired temperature. The tank supplying reactant B is dimensioned for the exact load. The batch loads are established such that it is impossible to reach the critical concentration of reactant B, even if the entire load of B was introduced with no reaction. In practice, for a critical concentration of B to be achieved, the operators must fill and empty the tank holding reactant B multiple times. Consequently, the advantage of the inherent safety of processes in batches is that it is more difficult to reach a hazardous condition. On the other hand, the drawback is that since the reactor volume is very large, the consequences of a breach of integrity from a potentially explosive reaction would also be serious.

A perfectly continuously stirred tank reactor (CSTR) is smaller, with a volume in the order of $1/10$ of the size of the discontinuous reactor, for the same production volume. Thus, the consequences of the potentially explosive reaction are much smaller if the critical concentration of B is reached. However, the continuous operation of the process involves simultaneously managing the feed stream of the mixture of reactant A in solution in the solvent S, the stream of reactant B and the supply of the catalyst C. With different ratios for the supply flow rate, it would be possible to produce almost any concentration of reactant B in the reactor. The reliability of the continuously operating installation depends on the instrumentation (flow-meters, ratio controllers, control valves, etc.) to ensure that the reactants are introduced in the correct ratio. Although the control to prevent a build-up of reactant B in the reactor could be considered redundant and reliable, it is difficult to exclude the possibility of failure. From this point of view, it follows that the process is inherently less safe. Thus, the continuous process offers both an inherent advantage, in terms of safety, as the reactor is much smaller, but also a drawback, as it depends on instrumentation and control to guarantee the supply of reactant B. At this stage, this comparison of the two processes demonstrates the difficulty of definitively claiming the concept of inherent safety.

Luyben and Hendershot (2004) also compared the behaviors of two plants, similar to those described at the beginning of this section, during the operation of a process for the nitration of an organic compound. The discontinuous semi-batch reactor, with a volume of $20\,m^3$, with a cooling jacket, is first loaded with the organic compound, dissolved in a solvent. The mixture of nitric acid and sulfuric acid is then gradually fed into the reactor at a flow rate that allows the exothermic reaction to be controlled. The management of this reactor implies vigilance for its operation. Indeed, if the nitric acid concentration exceeds 15%, the reaction mixture becomes very unstable and an explosion may occur. The acid feed comes in from a batch tank, which is sized such that it contains the exact quantity of acid required to react with the organic compound. Since the reaction kinetics are fast, in principle the nitric acid is consumed as it is fed in. The reactor does not normally contain unreacted nitric acid. However, if for any reason the reaction does not take place and the nitric acid concentration builds up in the reactor, the acid cannot exceed the explosivity limit of 15%, even if all the contents of the batch tank were added to the reaction tank. In terms of safety, it appears that the discontinuous process system is inherently safe from the point of view of the risk of an explosion caused by high concentration of nitric acid. This intrinsic protection does not a priori require any instrumentation. On the other hand, the intensification principle for a discontinuous process is not applicable, due to the large quantity of inflammable and corrosive products, the dispersion of which could pose a serious safety and environmental hazard in the case of a leak or loss of integrity of the reactor.

A CSTR, with a volume of 0.5 $m^3$, is fed by reactants through two lines, each of which includes a tank and a dosing pump. The reactor includes a cooling coil that occupies half the volume of the reactor, as it is important to remove the reaction heat. Consequently, the low liquid retention means that the quantity of the reactive mass could provoke smaller consequences in the case of a fire and toxic waste. However, it must be highlighted that the instrumentation governing the flow rate of the feed for the organic compound and acid must imperatively function. In a potential failure in the instrumentation, leading to the delivery of too much nitric acid, the reactive mix could quickly reach a concentration where it is unstable and explosive. The reliability of these instruments and control systems becomes a problem in the continuous reactor, which does not exist in the discontinuous batch systems. To sum it up, the intensified design of a continuously stirred reactor depends more on the reliability of the instrumentation and the control system, while the design of the discontinuous semi-batch process is inherently safer with regard to instrumentation.

This example illustrates a clear conflict between two inherently safer design strategies:

– The discontinuous semi-batch process is an example of the application of the "impact limitation" strategy for safety. The size of the nitric acid feed tank makes it impossible to reach the explosive concentration for the composition of the batch. The design also limits the possible impacts of a failure in controlling the nitric acid supply.

– The continuously stirred reactor is an example of an "intensified" process. The reactor is much smaller, which minimizes the consequences of a leak or a loss of integrity of the system.

#### 6.5.2.3. *Improving the manufacture of a mineral oxychloride*

Etchells (2005) gave a qualitative example of the improvement of a process to produce a mineral oxychloride. The process was initially used as a discontinuous process in a unit comprising six semi-batch reactors, with a 5 ton unit capacity. The production of mineral oxychloride was carried out through a reaction between chlorine and a mineral oxide. The large quantity of chlorine on the site constituted a significant hazard, despite the fact that in the case of failure, the chlorine supply would be cut off immediately (system shutdown). The process was carefully designed to avoid a build-up of the unreacted compounds. Since the reaction is exothermic, the reactors are cooled through immersed exchangers by water circulation. The risk of water leaking into the reactive mass could result in a violent reaction, hazardous to the integrity of the reactor capacities.

During a periodic check by competent external authorities, there was a demand to improve the overall safety of the unit. The operating company proposed replacing the semi-batch process with a continuous reactor operating on loop, producing 600 $m^3$ of product per week. The immediate stock of the reactive liquid was thus reduced by over two orders of magnitude. Further, the significant change in the chemistry of the reaction in favor of direct oxidation of chloride, removed the risk of handling large quantities of chloride in bulk, on the site. It must also be noted that the oxidation reaction is quick and exothermic, but that the actual rise in temperature in the reactor is smaller and remains less than 20°C. The product that is formed acts as a diluent for the reactants in the loop. Finally, an exterior, tubular thermal exchanger, equipped with water spray cooling systems at atmospheric pressure, now cools the reaction and eliminates the risk of water leaking into the reactor.

#### 6.5.2.4. *Effect of minimizing the liquid retention in the operation of a distillation column train*

Consider a direct distillation unit for a ternary mixture of benzene, toluene and o-xylene comprising two tray columns arranged in series. The first column, with a diameter of 2.2 m, is fed at a rate of 454 $kmole.h^{-1}$ with a mixture, whose composition is 40% mole fraction of benzene, 30% mole fraction of toluene and 30% mole fraction of o-xylene. This column, equipped with 30 theoretical trays, operates at a head pressure of 400 mmHg with a reflux of 1.6. The molar concentration of toluene in the distillate is 99% while the residue at the foot of the column has a 99% molar concentration of o-xylene. The "liquid retention" parameter must be carefully examined, as the intensification principle suggests that the *holdup* be low to minimize the effects of hazards due to the inflammability of the products.

However, low retention at the base of the column signifies that the flow rate disturbances in the column C1 are more rapid. A change in the supply to the unit will produce a more rapid variation in the flow rate at the bottom, exiting column C1. This rapid change in the supply to column C2 will produce a stronger disruption in its operation than a more gradual change. The best practices for selecting the components of a distillation column recommend considering a duration of 10 min. to completely fill the base of the column. The flow rate of the descent from the tray just above the base of column C1 is 1.6 $m^3.min^{-1}$. Consequently, under normal operation, the volume of the base of the column C1 is 16.4 $m^3$, while the retention volume is 8.2 $m^3$ to allow the disengagement of vapor bubbles. By adopting a time of 1 min., the same procedure in the intensified process leads to a liquid retention volume of 0.82 $m^3$.

Luyben and Hendershot (2004) simulated the behavior of the instrumented plant with two columns to test the influence of the liquid holdup. The feed flow rate to column C1 was subject to a positive step function of 20%. The details of the dynamic behavior of column C2, following the previous disruption, are described in their publication. To sum up, Table 6.9 provides certain comparison elements for the dynamic behavior in column C2 based on the liquid retention at the base of the column C1.

| Operating conditions | Normal plant | Intensified plant |
|---|---|---|
| Retention time C1 (min.) | 10 | 1 |
| Hold up volume base C1 ($m^3$) | 8.2 | 0.82 |
| Temperature difference of the sensitive tray C2 (°C) | –1.9 | –5.6 |
| Variation of the feed rate C2 | 20 % | 29 % |
| Concentration % toluene at the bottom of C2:<br>before step disturbance,<br>after step disturbance | 0.08<br>0.13 | 0.08<br>0.25 |

**Table 6.9.** *Comparison of the dynamic behavior of the C2 column as a function of the liquid retention in the base of the C1 column (Luyben and Hendershot 2004)*

To conclude with this example, while the application of the intensification principle leads to minimizing the liquid retention of inflammable products at the foot of column C1, the dynamic behavior of the plant seen in the simulation results clearly demonstrates the drawbacks of the intensification design in terms of operating stability and degradation in the quality of the products.

### 6.5.2.5. *Comparing two continuous processes for the nitration of an organic compound*

Luyben and Hendershot (2004) studied the functioning of two different continuous plants for the nitration of benzene to manufacture nitrobenzene. The first plant consisted of a continuously stirred reactor. The second was made up of two identical perfectly stirred reactors arranged in series. Each plant was continuously fed by two lines: benzene (flow rate 23 kmole.$h^{-1}$) and the 50% molar solution of nitric acid in water (flow rate 68 kmole.h$^{-1}$). The operating temperature is 121°C under a pressure of 1 bar. The required final conversion rate into benzene is 98%. Due to the highly exothermic nitration reaction, each reactor is equipped with a cooling jacket. Table 6.10 summarizes the respective operating parameters for the functioning of the plant containing a single reactor, and the first upstream reactor for the plant with two reactors in series.

| **Plant** | **One reactor** | **Two reactors (first reactor)** |
|---|---|---|
| Volume (m$^3$) | 122 | 14 |
| Diameter (m) | 4.3 | 2.1 |
| Height (m) | 8.5 | 4.2 |
| Conversion rate (%) | 98 | 87 |
| Jacket area (m$^2$) | 114 | 27 |
| Heat to be removed (kWatt) | 747 | 639 |
| Coolant flow rate (kg.h$^{-1}$) | 62,880 | 22,777 |
| Inlet temperature of the cold fluid (°C) | 107 | 79 |
| Outlet temperature of the cold fluid (°C) | 117 | 103.5 |

**Table 6.10.** *The respective operating parameters for a plant with a single reactor, and the first reactor in the plant with two reactors (Luyben and Hendershot 2004)*

A close examination of Table 6.10, leaving aside all consideration of the system dynamics, shows that the plant with two continuously stirred reactors is of more interest:

– there is a smaller inventory of hazardous material;

– the size of the equipment is smaller, which results in lower operating costs;

– the required flow rate of the cooling liquid is smaller for the same total heat elimination.

Luyben and Hendershot (2004) also used numerical simulation to study the dynamic behavior of two instrumented plants. Preliminary tests were carried out by subjecting the cooling fluid inflow to step functions. The plant with a single reactor produced a stable response, corresponding to a first order system. However, the plant with two reactors generated a decreasing, sub-damped oscillatory response. This observation indicates a more complex behavioral dynamic in the reactor containing a smaller volume of liquid. The authors also tested the influence of a disturbance in the benzene inlet flow rate. The disruption of the inflow consisted of an initial, incremental step function of 20% in the flow after 0.2 h of normal operation, followed by a second step function reducing the flow to 80% of its initial nominal value at the end of 5 h. The detailed observation of the dynamic behavior of the two plants are given in Luyben and Hendershot (2004). To sum up, the respective comparisons of the reactor temperature and flow rate of the cooling liquid clearly show that the plant with two reactors, with intensified design, presents a less stable behavioral dynamic. Despite its advantage in having an intensified design with a continuous regime, the plant with two reactors shows greater variability in behavior during disruptions in operational parameters.

#### 6.5.2.6. *Catalytic incineration of volatile organic compounds*

This example will compare two processing units for a process to destroy volatile organic compounds (VOCs) present in the gaseous emissions from a polymerization unit. The major VOC is glycol polyethylene. Each unit must treat a pollutant stream diluted in an inert flow of 5,000 $Nm^3.h^{-1}$ at an initial concentration of 0.001%. Figure 6.13 depicts the scheme showing the principle for the catalytic combustion process implemented in a fixed-bed reactor (Figure 6.13(a)) and in an intensified reactor (Figure 6.13(b)).

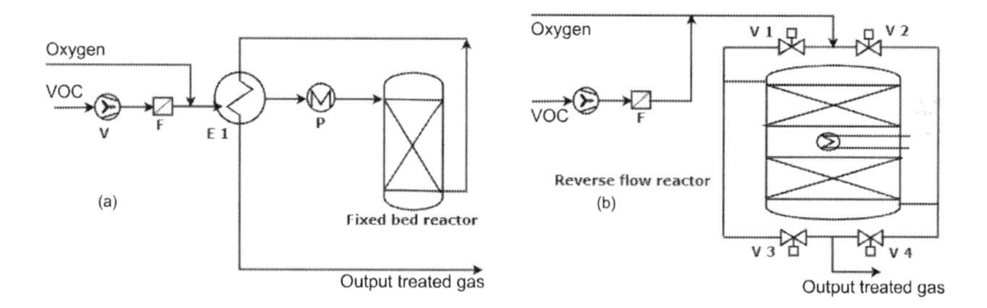


**Figure 6.13.** *Schemes showing the principle of the fixed-bed reactor (a) and the intensified reactor (b) (Baldissone et al. 2014)*

The conventional unit consists of a fixed-bed catalytic reactor. The gas flow to be treated is first filtered (F), then pre-heated in an exchanger by heat recovered from waste gas. It circulates in a descending current into the fixed-bed catalytic converter. Its instrumentation is described as simple, because it has only 10 alarms and a single automatic protection system.

The intensified unit is made up of a reverse-flow catalytic reactor, in which regenerative heat is recovered through the periodic inversion of the direction of the gas flow. The reactor is filled with successive layers packed with an inert solid and an active catalyst. Apart from a few elements that are similar to those in the fixed-bed installation, the intensified unit consists of a set of valves (V1, V2, V3, V4) used to periodically reverse the direction of gas stream. The operating principle for a reverse-flow reactor is especially simple. When the valves V1 and V4 are open (V2 and V3 are then closed), the gas stream, which flows from top to bottom, is headed by the inert solid layer located close to the reactor entry. It then undergoes slow, flameless catalytic combustion in the catalyst layers. Finally, it heats the inert solid layer provided at the bottom of the reactor, before exiting the reactor. After a certain time, which must be carefully chosen, the direction of flow is reversed by opening valves V2 and V3 (valves V1 and V4 are then closed). An electric heater is provided between the layers of the intensified reactor in order to maintain the catalyst temperature, if necessary. The instrumentation and control of this reactor, which are clearly more complex, include 30 alarms and 10 automatic protection systems.

Baldissone et al. (2014) studied the safety and reliability of these two plants. A recursive operability analysis (ROA) allowed them to identify the major top events and their possible causes:

- overpressure in the reactor;
- excessive VOC concentration in the output;
- catalyst sintering;
- release of the safety valve.

Excessive overpressure could cause a loss of integrity in the reactor. A large VOC concentration in the treated gas is the result of poor functioning of the catalytic combustion. Sintering of the catalyst is caused by uncontrolled temperature in the reactor, leading to the destruction of the catalyst. The release from the safety valve is due to excess pressure in the reactor.

The authors then carried out a quantitative analysis, using the fault tree analysis, to assess the probability of the top events. Table 6.11 presents the estimated respective values for each unit.

| **Top event in the reactor** | **Traditional plant** | **Intensified plant** |
|---|---|---|
| Reactor overpressure | $1.12 \times 10^{-3}$ | $6.75 \times 10^{-5}$ |
| Excess VOC concentration in output | $1.74 \times 10^{-2}$ | $6.57 \times 10^{-1}$ |
| Sintering catalyst | $5.70 \times 10^{-2}$ | $5.64 \times 10^{-5}$ |
| Release from safety valve | – | $4.80 \times 10^{-3}$ |

**Table 6.11.** *Probabilities of top events in each VOC treatment unit (Baldissone et al. 2014)*

Table 6.11 shows that a loss of integrity in the reactor has proven to be the least probable event. In this case, the intensified equipment then offers better results than the traditional plant. However, this result depends on the greater number of safety systems installed on the intensified equipment, which can intervene in the case of overpressure in the reactor. The top event related to excessive VOC concentration in the outflow is more probable than the earlier event. This condition can particularly affect intensified equipment due to its more complex control system for the output gas composition. The catalyst sintering event is more probable in a traditional plant. Indeed, the intensified reactor is equipped with a more sophisticated temperature control, limiting the risk of reaching the sintering temperature.

In summary, the results from the specific case study show that the heightened level of complexity of the intensified reactor requires the use of greater protection methods, as compared to the traditional unit. Thus, process intensification would appear to be less reliable, with regard to safety and reliability. However, this is an incomplete point of view, as it does not take into account the positive effects that intensification has on reducing the consequences of hazardous events, in terms of safety and reliability.

Baldissone et al. (2016) complemented the previous study with an integrated dynamic decision analysis, with the goal of evaluating the behavior of both plants in the case of failure in greater detail. These additional results show that the probability of the sintering of the catalyst is higher in the traditional plant, as the control system used in the reverse-flow reactor is more complex and has more redundancies. With regard to the VOC waste, this event is still more probable in the reverse-flow reactor than in the traditional plant; however, the associated risk is smaller, although the difference is not very large.

Finally, Baldissone et al. (2017) carried out a cost–benefit analysis to help select between competing technologies to process VOCs. The analysis of the results made it possible to conclude that, depending on the advantages that are relevant to the stakeholder, it is possible to draw the following conclusions:

– considering the loss of time following a breakdown, both alternatives seem comparable in terms of the consequences and the risks of VOC waste;

– for both units, the most critical system seems to be the oxygen input system and inspections and maintenance activities must be focused on these;

– when the advantages related to the minimizing of operational costs are taken into account, the intensified unit (reverse-flow reactor) is the best option, as it can function with a cheaper catalyst and requires less energy.

In conclusion, this example, which Baldissone et al. (2014) call an emblematic example, shows that when the considerations recounted here are taken into account, it is difficult to make an informed decision on the best technology to meet industrial safety needs, without a complete analysis. In this example, the reverse-flow reactor seems to still be the most efficient technology for treating the highly diluted VOC effluents released by the polymerization units.

### 6.5.3. *Vigilance when putting into practice the risk analysis methods based on the use of digital data*

Particular attention must be paid to quantitative risk analysis tools, based on the availability of digital data, not only for intensified processes but also, more broadly, for process safety, as they are relevant to the context of Industry 4.0. The proliferation of digital data and their generalized use must, however, be studied with scientific rigor to avoid applying these tools inappropriately.

Wen et al. (2022) have attempted to answer the following questions on the subject::

– What are the key data for analyzing process safety?

– What are the sources of the data?

– What does the data-based method represent?

– What are the most widespread myths and false beliefs about data-based methods?

– How frequent are these myths and false beliefs?

The authors thus analyzed a selection of 500 articles from open literature on process safety during the period from 1990 to 2020. The chief sources of the digital data listed were collected from experimental measurements (37%), results from investigations (25%) and expert opinions (24%). The key digital data from process safety analyses were then categorized into three categories, as shown in Table 6.12:

– data analysis;

– logical approach;

– data mining.

Table 6.12 also specifies the respective theory and focus for each category and details the various attributes of the corresponding methods founded on digital data.

By indicating that a conceptual error is considered as the application of a method or model, or the representation of digital data, without tracing back the reference to the scientific principles on which it is built, Wen et al. (2022) identified six flawed generic designs and 12 corresponding classes, as summarized in Table 6.13.

Wen et al. (2022) observed that 33% of the 500 articles analyzed, representing 288 examples, contained conceptual errors when using digital data. The two most frequently identified generic categories that resulted in errors are as follows:

– lack of attention to information about the data (163 examples);

– the use of Bayesian networks without formulating an appropriate hypothesis (58 examples).

| Category item | Data analysis | Logic driving | Data mining |
|---|---|---|---|
| **Theory** | Statistics | Logical graph | Machine learning |
| **Focus** | Surface of the data | Data projecting to the future | Inside of the data |
| **Data driving method** | Mean | Fault tree analysis | Artificial neural network |
| | Standard | Event tree | Support vector machine<br>Decision tree |
| | Distribution | Bow-tie | Random forest<br>Classification and regression tree |
| | Kurtosis | Bayesian network | Naïve Bayesian classification<br>$k$-means clustering |
| | Skewness | Fuzzy theory | $k$-nearest logistic regression |
| | Histogram | Analytical hierarchy process | Principal component analysis<br>Independent component analysis |
| | Control chart | Petri net | Partial least squares |

**Table 6.12.** *Data-driven methods in process safety analysis (Wen et al. 2022)*

In addition to these two categories, the other errors noticed were largely related to research based on machine learning tools and the use of multivariable statistical modeling of processes. The article uses various examples analyzing the details of incorrect and excessive applications and interpretations. Wen et al. (2022) further suggest various means through which these inappropriate applications can be avoided. In summary, the generalized use of risk analysis methods based on digital data requires a proper, detailed understanding of the limitations of these methods and a scientific approach to take into account the chemistry and physics underlying the processes.

| Myth/Misconception | Wrong class |
|---|---|
| Improper data representation | Digit inconsistency<br>Inaccurate computation to significant digits<br>False precision<br>Improper uncertainty |
| Absence of model behavior analysis for multivariable Statistical Process Monitoring | No examination of linear or non-linear data pattern<br>No identification on variable distribution |
| Missing Bayesian network's underlying assumption | Misuse of the overall dependence modeling<br>Unclear conditional probability tables |
| Overuse of artificial neural network | Using artificial neural networks to replace simple analytical equations<br>Setting arbitrary hyper parameters |
| Using correlation coefficient for model verification | Using correlation coefficient for model verification |
| Absence of error analysis | Absence of error analysis |

**Table 6.13.** *Commonly noticed myths and misconceptions (Wen et al. 2022)*

# Conclusion

This book offers a new and original way to raise awareness among actors intervening in safety to the problems brought in by the digital industrial revolution. Controlling risks in plants and factory units of the future is indeed a requirement today for society as a whole, especially for the stakeholders. Disruptive digital technology or innovation in the fields of communication, interconnection and data management in Industry 4.0 introduce new risks. The analysis of these risks reveals new approaches to be taken, or, at the very least, the updating of the best practices being applied. The new concept of Safety 4.0 involves a proactive change in the science of industrial safety, both for process safety as well as occupational safety and health. This change focuses on the resilience of systems and dynamic risk management. However, putting in place a procedure to satisfy these conditions is not a trivial task.

An important first consideration is to re-examine and specify the components of the safety management systems. The primordial issue is whether or not there is a fundamental difference between process safety and occupational safety and health.

Let us recall that process safety consists of managing all systems and units being operated by applying inherently safer design principles, flexible engineering, and rigorous best practices for operation. It deals with preventing, mitigating, attenuating and protecting against incidents and accidents that could lead to a hazardous loss of control of materials or energy. This kind of loss of control could provoke severe consequences such as a fire, explosion and/or toxic effects, finally leading to considerable material damages, a severe impact on the environment and stoppage or reduction in production, with financial losses and damage to the brand image and reputation. Breaches in process safety do not necessarily harm the concerned human operators.

On the other hand, occupational safety and health (OSH) is a multidisciplinary approach that aims to remove or reduce the risk of accidents that could occur when occupational activities are carried out. Occupational hazards could result from poorly controlled activity, restrictive postures or the use of toxic chemical products. The broad categories of occupational hazards include risks related to physical activity, risks of falls and slips, risks of infection, psychosocial risks, risks related to atypical work timings and risks related to movements. These risks could lead to bodily and/or psychological harm and occupational diseases with immediate or delayed effects.

The study of the statement and management of the balance between the two types of safety is essential. For example, Hopkins (2009) and Grote (2012) analyzed the return of experience from the enquiry conducted by the US Chemical Safety and Hazard Investigation Board (CSB) regarding the explosion at the British Petroleum (BP) plant in Texas City in 2005. It was observed that the BP upper management had emphasized personnel safety and rewarded supervisors who had obtained low rates of occupational hazards within their teams. However, at the same time there were many flagrant warning signs related to dips in process safety that had not been taken into account. This case is a good illustration of the fact that increasing the overall safety level of a plant requires paying particular attention to both process safety and individual safety in the workplace.

The study of certain characteristics of the respective roles played by process safety and occupational safety showed an overlap between some elements, such as the analysis safety systems, risk assessments, risk management, and human factors. Considering the challenges posed by interconnections, autonomous equipment and the combined automation of human behavior and supervisory monitoring, the result of the overlap made it possible to establish an inventory of the risk-assessment and risk-analysis methods and techniques. Their concepts, paradigms, bases for structuring, corresponding coupling scores and complexity have been systematically defined. Consequently, conventional methods such as the preliminary risk analysis, the HAZOP method, fault tree analysis, bow-tie graphs, LOPA method and probabilistic graphical models with Bayesian networks are still usable. The HAZOP-LIKE method, with its specific key words, is well adapted to the risk analysis of microstructured equipment. Finally, the prevalence of integrated systemic methods of holistic and sociotechnical systems is recommended. The complex Safety 4.0 methods and techniques that respond to Industry 4.0 requirements of interconnection and complexity are the EAST, FRAMP and STAMP methods. The many examples presented in Chapters 5 and 6 in this book illustrate the various methodological approaches for Safety 4.0.

A second consideration is the reasoned and simultaneous use of performance indicators. The two families of performance indicators, lagging indicators and leading indicators, must be used simultaneously. Let us recall that the lagging indicator is said to be an a posteriori indicator, while the leading indicator is a proactive indicator. Many companies chiefly publish and share the values of lagging indicators in the field of occupational safety and health. This does not give any information on how well-controlled process safety is. It is imperative to use leading indicators to promote an effective risk management strategy. The book indicates several recommendation guides to indicate lagging and leading indicators (OECD 2008; Rogers et al. 2009; CCPS 2010; HSE 2010; Mazri 2015; OSHA 2019; UICh 2017). Finally, reciprocal synergy between lagging and leading indicators is often helpful in improving the creation of a safety management system.

At this stage, the potential application of Safety 4.0 in the industry of the future through the deployment of digitization, artificial intelligence, using Big Data, the popularization of virtual reality and digital twins should allow a significant leap in modeling, monitoring and prediction technologies. The significant progress expected from the digital revolution should enhance the knowledge in an increasingly complex world, despite the limitations and uncertainties that are either poorly considered or not taken into account at all. The feasibility of completely controlling safety in industrial systems is still limited. Thus, the risk of considering process safety to be inherently safer, a priori, even for plants constructed using inherently safer design, remains a clearly ambiguous one. Several examples given in Chapter 6 highlighted conflicts between Safety 4.0 and the process, especially when studying the dynamic behavior of intensification processes.

A third and important concern for Industry 4.0 is related to prioritizing the availability of a pool of competent operators, technicians and engineers who have mastered the consequences of digital technologies in implementing Safety 4.0. Amalberti (2021) analyzed foreseeable evolutions across generations of employers in the period from 2030 to 2040 and emphasized the joint presence of new, young hires and experienced employees forced to push the age of retirement. What are the skills that would then be required for Safety 4.0?

Restricting ourselves only to the actors in the industrial domains of chemistry, chemical engineering and process engineering, the first absolute need is to bring together the currently disparate jargon used by this population and that used by experts in artificial intelligence and digitization in order to have mutual and reciprocal understanding (André 2021). Next, the framework for changes in the basis training and training through learning for new generations must be revised and adapted to the requirements of Safety 4.0. Keller (2018) published the results of a study into the potential needs in terms of digital technologies in different sectors of activity in the German chemical industry.

Table C.1 offers an example, ranking digital applications in the sector of industrial engineering on a scale of 0 to 10, based on their relevance in 2025 and the development between 2018 and 2025. Studying this table shows us that one of the expectations of Industry 4.0 deals above all with cloud storage, big data management, virtual and augmented reality, and, of course, the modeling and simulation already widely used in industrial engineering. Since this study, Khan et al. (2021) have proposed syllabi for both bachelor's and master's programs integrating the basics of digitalization with respect to managing degraded situations, the control and automation of processes, the reliability of processes and the integrity of the equipment. Gajek et al. (2022) suggested two approaches for training employees 4.0. The first, a hybrid approach, consisted of combining a conventional process safety course with a choice of one specialized digital field. The second suggestion, aimed at the younger generation that is more familiar with modern technology, is an integrated Safety 4.0 and Industry 4.0 training. Laurent and Fabiano (2022) listed the requirements for the new skills required for Industry 4.0 conditions.

For example, the large-scale recruitment of data specialists (data engineers, data scientists) that is required is disrupted by the lack of clarity on the ethics of the functions to be carried out. These authors have proposed a vision of new trends in Education 4.0 and listed the orientations that could be applied in educating a process Safety 4.0 workforce.

To what extent are current job posts likely to be affected by these digital evolutions?

| | Blockchain | Robotics | Manufacturing 3D | Machine learning | Deep learning | IoT | Cloud | Big Data | AR VR | Modeling Simulation |
|---|---|---|---|---|---|---|---|---|---|---|
| Pertinence 2025 | 3.5 | 4.2 | 5.1 | 5.3 | 5.2 | 6.1 | 6.4 | 6.8 | 6.9 | 7.9 |
| Evolution 2018-2025 | 5.1 | 2.1 | 3 | 2.6 | 2.5 | 2.6 | 2.6 | 2.5 | 2.1 | 1.5 |

**Table C.1.** *The relevance of digital technologies in 2025 for industrial engineering and the factor of evolution from 2018-2025 (Keller 2018)*

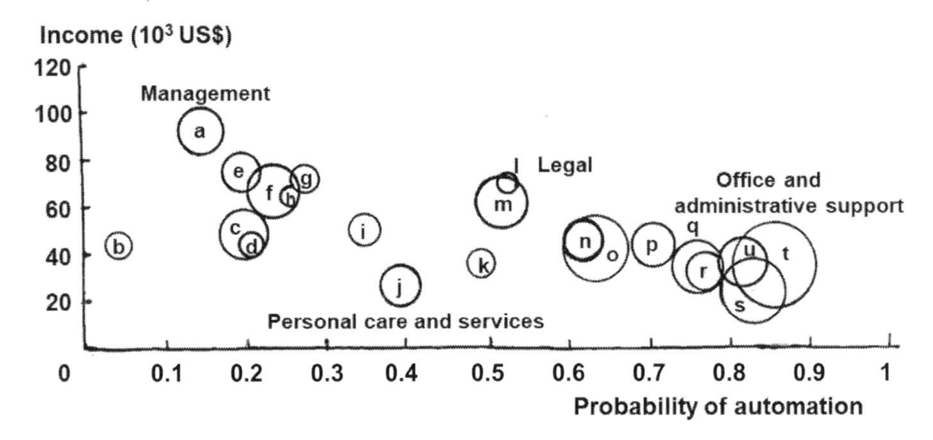


**Figure C.1.** *Relationship between the annual incomes of different professional sectors and the probability of their being automated. Reference year – 2017 (Cancelada 2019)*

REMARK ON FIGURE C.1.– *(a) Management; (b) community and social services; (c) education, training and library; (d) art, design, entertainment, sport and media; (e) computer science and mathematics; (f) health care practitioners and technicians; (g) architecture and engineering; (h) life, physical and social sciences; (i) protective services; (j) personal care and services; (k) health care support; (l) legal; (m) business and financial transactions; (n) installation, maintenance and repairs; (o) sales and service activities; (p) construction and extraction; (q) transport and movement of materials; (r) agriculture, fishing and forestry; (s) food products and related services; (t) offices and administrative support; (u) production.*

Cancelada (2019) aggregated data from 702 professions, across different sectors of activity in the American labor market, compiled by Frey and Osborne (2017) to study the impacts of digitization. For example, he studied how automation could affect revenues. The bubble graph in Figure C.1 demonstrates the relationship between annual incomes for different occupational sectors and the probability that they would be automated.

Specific probabilities for different occupations can be found in Appendix A in the article by Frey and Osborne (2017). Considering three scenarios for the employees whose jobs would be automated, Cancelada noted that the most realistic scenario would result in a 20% dip in the initial revenues. It is, however, useful to highlight the need for caution in interpreting these numbers, which are based on very specific hypotheses that may be influenced by other factors in the future. For example, this reported analysis does not take into account how the job market will adapt to automation. There will be changes not only in salary but also in the employment flows between industries, and the creation of new industries and so on.

Finally, it is the responsibility of all stakeholders from universities, in the industry and centers of learning to advance the various contributions identified from the Safety 4.0 approach for industries of the future.

# References

Abdo, H. (2017). Dealing with uncertainties in risk analysis: Combining safety and cybersecurity. PhD Thesis, Université Grenoble Alpes, Grenoble.

Abdo, H. and Flaus, J.M. (2015). A mixed fuzzy probabilistic approach for risk assessment of dynamic systems. *Science Direct IFAC*, 48(3), 960–965.

Abdo, H., Kaouk, M., Flaus, J.M., Masse, F. (2017). Towards a better industrial risk analysis: A new approach that combines cybersecurity within safety. *Archives HAL-Ineris*, 0185345, 1–8 [Online]. Available at: https://hal-ineris.archives-ouvertes.fr/ineris-01853454.

Abdo, H., Kaouk, M., Flaus, J.M., Masse, F. (2018). A safety/security risk analysis approach of Industrial Control Systems: A cyber bowtie combining new version of attack tree bowtie analysis. *Computers and Security*, 72, 175–195.

Adriaensen, A., Decre, W., Pintelan, L. (2019). Can complexity thinking methods contribute to improving Occupational Safety in industry 4.0? A review of safety analysis methods and their concepts. *Safety*, 5, 1–33.

Agnusdei, G.P., Ela, V., Gnoni, M.G. (2021). A classification proposal of digital twin applications in the safety domain. *Computers and Industrial Engineering*, 154, 107–137.

Alberts, C.J., Sandra, G., Behrens, R.D. (1999). Operational Critical Threat Asset and Vulnerability Evaluation (OCTAVE) framework. Technical report, University of Pittsburgh.

Allford, L. and Wood, H. (2021). Accident analysis benchmarking exercise. Technical report, Union européenne, Luxembourg.

Almajali, S., Dhiah el Diehn, I.A.T., Haytem, S.B., Moussa, A. (2019). A distributed multilayer MEC cloud architecture for processing large scale IoT based multimedia applications. *Multimedia Tools and Applications*, 78(17), 24617–24638.

Amalberti, R. (2021). Risques émergents : la sécurité du futur, la sécurité au travail se réinvente. *YouTube* [Online]. Available at: https://www.youtube.com/watch?v=ubSsZwq276k.

Amyotte, P., Khan, F., Kletz, T. (2009). Inherently safer design activities over the past decade. In *Symposium Series No. 155 – Hazards XXI*. IChemE, Manchester.

André, J.C. (2019). *Industry 4.0: Paradoxes and Conflicts*. ISTE Ltd, London, and John Wiley & Sons, New York.

André, J.C. (2021). L'Intelligence artificielle au service du génie des procédés. *YouTube* [Online]. Available at: https://www.youtube.com/watch?v=DDr2OT_AczQ.

ANSSI (2014). *La cybersécurité des systèmes industriels – Méthodes de classification et mesures principales*. Agence nationale de la sécurité des systèmes d'information, Paris.

Anteol, B., Guillemy, N., Leroy, A. (2004). *Évaluation des risques professionnels – Questions réponses sur le document unique ED 887*. INRS, Paris.

Arab, S., Commenge, J.M., Portha, J.F., Falk, L. (2014). Methanol synthesis from carbon dioxide and hydrogen in multi-tubular fixed bed reactor and in multi-tubular reactor filled with monoliths. *Chemical Engineering Research and Design*, 92, 2598–2608.

Aubertin, G. (2007). *Management de la santé et de la sécurité au travail – Construire vos indicateurs pour atteindre vos objectifs ED 6013*. INRS, Paris.

Aubertin, G., Drais, E., Favaro, M., Meyer, A. (2007). Gestion des risques professionnels. *Techniques de l'Ingénieur*, SE 3910, 1–17.

Badri, A., Boudreau-Trudel, B., Souissi, A.S. (2018). Occupationnal health and safety in the industry 4.0 era: A cause for major concern? *Safety Science*, 109, 403–411.

Baldissone, G., Cavaglia, G., Demichela, M. (2014). Are intensified processes safer and more reliable than traditional processes? An emblematic case study. *Chemical Engineering Transactions*, 36, 415–420.

Baldissone, G., Fissore, D., Demichela, M. (2016). Catalytic after-treatment of lean VOC-air treatment: Process intensification vs. plant reliability. *Process Safety and Environmental Protection*, 100, 208–219.

Baldissone, G., Demichela, M., Fissore, D. (2017). Lean VOC-air mixtures catalytic treatment: Cost-benefit analysis of competing technologies. *Environments*, 4(46), 1–19.

Baybutt, P. (2002). Layers of protection analysis for human factors (LOPA-HP). *Process Safety Progress*, 21(2), 119–124.

Bayer, T., Jenck, J., Matlosz, M. (2005). IMPULSE – A new approach to process design. *Chemical Engineering and Technology*, 28(4), 431–438.

Becht, S., Franke, R., Geisselmann, A., Hahn, H. (2007). Micro process technology as a means of process intensification. *Chemical Engineering and Technology*, 30(3), 295–299.

Becht, S., Franke, R., Geisselmann, A., Hahn, H. (2009). An industrial view of process intensification. *Chemical Engineering and Processing: Process Intensification*, 48, 329–332.

Behesti, B.K., Chi, H.L., Wang, X. (2015). Using RFID systems in design and construction of chemical plants. In *The 9th International Chemical Engineering Congress and Exhibition*. IChEC, Shiraz.

Behr, A., Ebbers, W., Wiese, N. (2000). Miniplants – Ein Beitrag zur inhärenten Sicherheit? *Chemie Ingenieur Technik*, 72(10), 1157–1166.

Benaissa, W. (2007). Développement d'une méthodologie pour la conduite en sécurité d'un réacteur continu intensifié. PhD Thesis, LGC, CNRS, ENSIACET, INPT, Toulouse.

Benaissa, W., Elgue, S., Gabas, N., Carson, D., Demissy, M. (2005). Transposition of an exothermic reaction to a continuous intensified reactor. In *6th International Conference on Process Intensification*. BHR Group Limited, Bedfordshire/Delft.

Benaissa, W., Carson, D., Demissy, M. (2008a). Développement d'une méthodologie pour la conduite en sécurité d'un réacteur continu et intensifié. Technical report, DRA INERIS, Verneuil-en-Halatte.

Benaissa, W., Gabas, N., Cabassud, M., Carson, D., Elgue, S., Demissy, M. (2008b). Evaluation of an intensified continuous heat-exchanger reactor for inherently safer characteristics. *Journal of Loss Prevention in the Process Industries*, 2(5), 528–536.

Bevilacqua, M., Bottani, E., Ciarapica, F.E., Costantino, F., Donato, L.D., Ferraro, A., Mazzuto, G., Monteriu, A., Nardini, G., Ortenzi, M. et al. (2020). Digital twin reference model development to prevent operators' risk in process plants. *Sustainability*, 12(1088), 1–17.

Blaise, J.C., Brun, L., Savescu, A., Sghaier, A., Tihany, D., Wioland, L. (1993). *10 questions sur les robots collaboratifs – ED 6386*. INRS, Vandoeuvre-lès-Nancy.

Breque, M., De Nul, L., Petridis, A. (2021). *Industry 5.0 – Towards a Sustainable, Human-centric and Resilient European Industry*, volume R and I, 1st edition. European Commission, Directorate-General for Research and Innovation, Publications Office [Online]. Available at: https://data.europa.eu/doi/10.2777/308407.

Cabreira, C., Capistrano, M., Pacifio, A., Alves, M., Zacarin, P., Andreo, A., Albuquerque, H. (2015). RFID applied to protective equipment inspection. In *Brazil RFID*. IEEE, Sao Paulo.

Cancelada, G. (2019). Workplace automation: Should we fear the robots? [Online]. Available at: https://www.stlouisfed.org/open-vault/2019/october/workplace-automation-should-we-fear-robots.

Carlo, F.D., Mazzuto, G., Bevilacqua, M., Ciarapica, F.E. (2021). Retrofitting a process plant in an industry 4.0 perspective for improving safety and maintenance performance. *Sustainability*, 13(646), 1–18.

Carvalho, A., Gani, R., Matos, H. (2008). Design of sustainable chemical processes: Systematic retrofit analysis generation and evaluation of alternatives. *Process Safety and Environmental Protection*, 86, 328–346.

CCPS (1996). *Inherently Safer Chemical Processes: A Life Cycle Approach*. Wiley, New York.

CCPS (2010). *CCPS Process Safety Leading and Lagging Metrics*. CCPS AIChE, New York.

CCPS (2012). *Guidelines for Engineering Design for Process Safety*. CCPS AIChE, New York.

Charpentier, J.C. (2002). The triplet molecular process–product–process engineering: The future of chemical engineering? *Chemical Engineering Science*, 57, 4667–4690.

Charpentier, J.C. (2005). Review – Process intensification by miniaturization. *Chemical Engineering and Technology*, 28(3), 255–258.

Charpentier, J.C. (2007). In the frame of globalization and sustainability, process intensification, a path to the future of chemical and process engineering (molecules into money). *The Chemical Engineering Journal*, 134(1–3), 84–92.

Charpentier, J.C. (2016). Intensification des procédés – Introduction. *Techniques de l'Ingénieur*, J 7000, 1–6.

Chatouane, F. (2015). An overview on RFID technology and application. *IFAC*, 48(3), 382–387.

Chaudary, M. and Chopra, A. (2017). *CMMI for Development*. Springer, New York.

Chen, G., Yue, J., Yuan, Q. (2008). Gas-liquid microreaction technology: Recent developments and future challenges. *Chinese Journal of Chemical Engineering*, 16(5), 663–669.

Commenge, J.M. (2020). Big Data et intelligence artificielle pour le génie des procédés. École d'ingénieur. HAL, 03107557, 1–128 [Online]. Available at: https://hal.univ-lorraine.fr/hal-03107557.

Commenge, J.M. and Falk, L. (2014). Methodological framework for choice of intensified equipment and development of innovative technologies. *Chemical Engineering and Processing: Process Intensification*, 84, 109–127.

Commenge, J.M., Falk, L., Corriou, J.P., Matlosz, M. (2004). Intensification des procédés par microstructuration. *Comptes Rendus Physique*, 5, 597–608.

Commenge, J.M., Falk, L., Corriou, J.P., Matlosz, M. (2005). Analysis of microstructured reactor characteristics for process miniaturization and intensification. *Chemical Engineering and Technology*, 28(4), 446–458.

Cormier, A. and Ng, C. (2020). Integrating cybersecurity in hazard and risk analyses. *Journal of Loss Prevention in the Process Industries*, 64, 104044–104049.

Corrigan, T.E. and Stichweh, L.A. (1968). Esterification process development. *Chemical Engineering Science*, 23, 991–1002.

Cox, L.A.J. (2008). What's wrong with risk matrices? *Risk Analysis*, 28(2), 497–512.

Dallat, C., Salmon, P., Goode, N. (2019). Risky systems versus risky people: To what extent do risk assessment methods consider the systems approach to accident causation? A review of the literature. *Safety Science*, 119, 266–279.

Daniellou, F. and Descazeaux, M. (2019). De la prévention des accidents majeurs graves, mortels et technologiques majeurs. *L'essentiel – ICSI*, 1–24.

Debray, B. (2006). Méthodes d'analyse des risques générés par une installation industrielle. Technical report, INERIS, Verneuil-en-Halatte.

Delatour, G., Laclemence, P., Calcei, D., Mazri, C. (2014). Safety performance indicators: A questioning diversity. In *6th Italian Conference on Safety and Environmental in Process and Power Industry (CISAP 6)*. AIDIC, Bologne.

Delvosalle, C., Fiévez, C., Pipart, A., Debray, B. (2006). ARAMIS project: A comprehensive methodology for the identification of reference accident scenarios in process industries. *Journal of Hazardous Materials*, 130(3), 200–219.

Deur, J. (2020). Filière cuir au Bangladesh. Technical report, French embassy in Bangladesh.

Devries, W. (2007). European roadmap for process intensification. Technical report, Ministry of Economic Affairs [Online]. Available at: www.efce.info/efce_media/European_Roadmap_PI- p-531.pdf.

Dixit, S., Yadav, A., Dwidedi, P.D., Das, M. (2015). Toxic hazards of leather industry and technologies to combat threat: A review. *Journal of Cleaner Production*, 87, 39–49.

Drais, E. (2018). Le management de la santé et sécurité (SST) : levier essentiel d'une culture de prévention. *Cahiers de notes documentaires, Hygiène et sécurité du travail*, 253, 22–51.

DRT (2006). Instruction DRT du 14/04/06 relative à la collaboration renforcée entre les inspections chargées du contrôle des établissements classés "Séveso seuil haut". *Aida* [Online]. Available at: https://aida.ineris.fr/reglementation/instruction-drt-140406-relative-a-collaboration-renforcee-entre-inspections-chargees.

Duval, C., Leger, A., Weber, P., Levrat, E., Farret, R. (2007). Choice of a risk analysis method for complex socio-technical systems. In *18th European Safety and Reliability Conference*. ESREL, Stavanger.

EBIOS (2010). *Guide méthodologique : méthode de gestion des risques*, 1st edition. Agence nationale de la sécurité des systèmes d'information, Paris.

EBIOS (2018). *Guide méthodologique : EBIOS Risk Manager*, 1st edition. Agence nationale de la sécurité des systèmes d'information, Paris.

EBIOS (2019). *Le supplément : les fiches méthodes EBIOS Risk Manager*, 1st edition. Agence nationale de la sécurité des systèmes d'information, Paris.

Englund, S.M. (1982a). Chemical processing: Batch or continuous, part 1. *J. Chemical Education*, 59(9), 766–768.

Englund, S.M. (1982b). Chemical processing: Batch or continuous, part 2. *J. Chemical Education*, 59(10), 860–862.

Englund, S.M. (1990). *Opportunities in the Design of Inherently Safer Chemical Plants*, volume 15. Academic Press, New York.

Englund, S.M. (1991). Design and operate plants for inherent safety, part 1. *Chem. Eng. Progress*, 85–91.

Englund, S.M. (1995). Inherently safer plants: Practical applications. *Process Safety Progress*, 14(1), 63–70.

EPSC (2021). Process safety fundamentals – Safe operational principles to avoid incidents with hazardous chemicals. *EPSC Web Seminar, European Process Safety Centre* [Online]. Available at: https://epsc.be.

Etchells, J.C. (2005). Process intensification – Safety pros and cons. *Process Safety and Environmental/Protection*, 83(B2), 85–89.

Fabbri, L. and Contini, S. (2009). Benchmarking on the evaluation of major accident related risk assessment. *Journal of Hazardous Materials*, 162, 1465–1476.

Falk, L., Porta, J.F., Commenge, J.M. (2019). Theoretical principles of flow chemistry. *Techniques de l'Ingénieur*, J 8025, 1–24.

Flaus, J.M. (2019). *Cybersecurity of Industrial Systems*. ISTE Ltd, London, and John Wiley & Sons, New York.

Florent, M., Commenge, J.M., Falk, L., Sebastien, L. (2013). Technologies comparison for iterative data acquisition strategies. *Chemical Engineering Science*, 109, 829–838.

Forcina, A. and Falcone, D. (2021). The role of industry 4.0 enabling technologies for safety management: A systematic literature review. In *Procedia Computer Science – International Conference on Industry 4.0 and Smart Manufacturing*. ISM, Hagenberg.

Frey, C.B. and Osborne, M.A. (2017). The future of employement: How susceptible are jobs to computerisation? *Technological Forecasting and Social Change*, 114, 254–280.

Gajek, A., Fabiano, B., Laurent, A., Jensen, N. (2022). Process safety education of future employee 4.0 in industry 4.0. *Journal of Loss Prevention in the Process Industries*, 64(104696), 1–29.

Gourdon, C. (2016). Intensification des procédés – Fondamentaux et exemples d'industrialisation. *Techniques de l'Ingénieur*, J 7002, 1–20.

Gourdon, C., Elgue, S., Prat, L. (2018). What are the needs for process intensification? *IFP Énergies Nouvelles*, 70(3), 463–473.

Gowland, R. (1996). Putting numbers on inherent safety. *Chemical Engineering*, 103, 82–86.

Gowland, R. (2006). The accidental risk assessment methodology for industries (ARAMIS) – Layer of protection analysis (LOPA) methodology: A step forward towards convergent practices in risk assessment? *Journal of Hazardous Materials*, 130, 307–310.

de Graaf, R.A. and Tikku, R. (2007). Inherently safe reactor design using microreactor. IChemE Symposium Series No. 153. In *12th International Symposium Loss Prevention and Safety Promotion in the Process Industries*. Edinburgh.

Grote, G. (2012). Safety management in different high-risk domains. All the same? *Safety Science*, 50, 1983–1992.

Guan, J., Graham, J.H., Hieb, J.L. (2011). A digraph model for risk identification and management in SCADA system. In *International Conference of Intelligence and Security Informatics*. IEEE Computer Society, Beijing.

Hankel, M. and Rexroth, B. (2015). The reference architectural model industrie 4.0 (rami 4.0). *Zentral Verband Electrotechnik und Electron Industrie (ZVEI)*.

Hardy, K. (2010). Contribution à l'étude d'un modèle d'accident systémique, le cas du modèle STAMP : application et pistes d'amélioration. PhD Thesis, École nationale supérieure des Mines de Paris, Paris.

Hardy, K. and Guarnieri, F. (2013). Hazard mitigation through a systemic model of accident to a socio-technical system: A case study. *Journal of Energy and Power Engineering*, 7(4), 775–787.

Hassim, M., Hurme, M., Kidam, K. (2010). Inherent health consideration for workers'protection in chemical plants. *Chemical Engineering Transactions*, 19, 353–358.

Heikkila, A.M. (1999). Inherent safety in process plant design – An index based approach. PhD Thesis, Helsinki University of Technology/VTT Technical Research Centre of Finland, Helsinki.

Heikkila, A.M., Hurme, M., Jarvelainen, M. (1996). Safety consideration in process synthesis. *Computers and Chemical Engineering*, 20(Supplement A), 115–120.

Helmus, M. (2007). Application field of RFID in health safety and environmental management. In *1st Annual RFID Eurasia*. IEEE, Istanbul.

Hendershot, D.H. (1995). Conflicts and decisions in the search for inherently safer process options. *Process Safety Progress*, 14(1), 52–56.

Hendershot, D.H. (2000). Process minimization: Making plants safer. *Chem. Engng. Progress*, 96(1), 35–40.

Hermann, M., Pentek, T., Otto, B. (2016). Design principles for Industry 4.0 scenarios. In *49th Hawaï International Conference on System Sciences*. IEEE Computer Society, Kōloa.

Hessel, V. (2009), Review: Novel process windows – Gate to maximizing process intensification via flow chemistry. *Chemical Engineerig and Technology*, 32(11), 1655–1681.

Hessel, V. and Lowe, H. (2005). Manufacturing microreactor technology: Applications in pharma/chemical processing. *Innovations in Pharmaceutical Technology*, 88–92.

Hessel, V., Cortese, B., de Croon, M.H.J.M. (2011). Novel process windows – Concept, proposition and evaluation methodology, and intensified superheated processing. *Chemical Engineerig Science*, 66(7), 1426–1448.

Hessel, V., Kralisch, D., Kockmann, N., Noel, T., Wang, Q. (2013). Novel process windows for enabling, accelerating and uplifting flow chemistry. *ChemPubSoc Europe*, 6, 746–789.

Hettinger, L., Kirlik, A., Gob, Y.M., Buckle, P. (2015). Modeling and simulation of complex sociotechnical systems: Envisioning and analysing work environment. *Ergonomics*, 58(4), 600–614.

Himmelblau, D.M. (2000). Applications of artificial neural networks in chemical engineering. *Korean Journal of Chemical Engineering*, 17(4), 373–392.

Hoguin, S. (2018). Santé – Sécurité au travail : les points clés à connaître sur l'ISO 45001. *Techniques de l'Ingénieur*, Innovations sectorielles, 1–6.

Hollnagel, E. and Speziali, J. (2008). Study on developments in accident investigation methods: A survey of the state of the art. Technical report, SKI, Stockholm.

Hopkins, A. (2009). Thinking about process indicators. *Safety Science*, 47, 460–465.

Hourtoulou, D. and Salvi, O. (2003). ARAMIS: Accidental risk assessment methodology for industries in the framework of Seveso 2 directive. *Préventique*, 69, 47–49.

HSE (2010). *Developing Process Safety Indicators*, 1st edition. HSE, London.

ICCA (2017). Guidance for reporting on the ICCA globally harmonized process safety metric, part 2. Technical document, ICCA, Paris.

Iddir, O. (2008). Le nœud papillon : une méthode de quantification du risque majeur. *Techniques de l'Ingénieur*, SE 4055, 1–23.

Iddir, O. (2021). Évolutions de la méthode LOPA. *Techniques de l'Ingénieur*, ESE 4076, 1–23.

INSET (2001). The inherent safety health environmental evaluation tool (INSIDE project). Report, European Union, Brussels.

ISO (2018). Norme 45001 – Systèmes de management de la santé et de la sécurité au travail – Exigences et lignes directrices pour leur utilisation, 1st edition. Organisation internationale de normalisation, Geneva.

ISO (2019). Norme 31010 – Management du risque – Techniques d'appréciation du risque, 2nd edition. Organisation internationale de normalisation, Geneva.

Jahnisch, K., Baerns, M., Hessel, V., Ehrfeld, W., Haverkamp, V., Lowe, H., Wille, C., Guber, A. (2000). Direct fluorination of toluene using elemental fluorine in gas-liquid microreactors. *Journal of Fluorine Chemistry*, 105, 117–128.

Jimenez, M. (2017). What is the impact of the industry 4.0 in the process industry? Master's degree, Delft University of Technology, Delft.

Jones, S. (2019). Managing process safety in the age of digital transformation. *Chemical Engineering Transactions*, 77, 619–624.

Jones, S. and Menon, A. (2021). Visualize and manage process safety risk operation – Practical case studies. EPSC/SPHERA [Online]. Available at: https://epsc.be

Julien, N. and Martin, E. (2018). *L'usine du futur – Stratégies et déploiement – Industries 4.0, de l'IoT aux jumeaux numériques*. Dunod, Malakoff.

Kagermann, H., Wahlster, W., Helbie, J. (2013). Recommandations for implementing the strategic initiative Industry 4.0. Technical report, Acatech – National Academy of Science and Engineering, Francfort-sur-le-Main.

Keller, W. (2018). Berufe 4.0 – Wie Chemiker und Ingenieure in der digitalen Chemie arbeiten. Technical report, VCW/GDCh, Francfort-sur-le-Main.

Kellezi, D., Boegelund, C., Meng, W. (2019). Towards secure open banking architecture: An evaluation with OWASP. In *International Conference on Network and System Security*. Springer, New York.

Kerin, T. (2019). *Managing Process Safety – Core Body of Knowledge for the Generalist OHS Professional*. AIHS, Tullamarine.

Khakzad, N., Khan, F., Amyotte, P. (2013). Dynamic safety analysis of process systems by mapping bow-tie into Bayesian network. *Process Safety and Environmental Protection*, 91, 46–53.

Khan, F. and Amyotte, P. (2003). How to make inherent safety practice a reality? *The Canadian Journal of Chemical Engineering*, 81, 2–16.

Khan, F., Amyotte, P., Adedigba, S. (2021). Process safety concerns in process system digitalization. *Education for Chemical Engineers*, 34, 33–46.

Kidam, K., Hurne, M., Hassim, M.H. (2010). Technical analysis of accidents in chemical process industry and lessons learnt. In *4th International Conference on Safety and Environment in the Process Industry*. AIDIC, Florence.

Klais, O., Westphal, F., Benaissa, W., Carson, D. (2009a). Guidance on safety/health for process intensification including MS design, part i: Reaction hazards. *Chemical Engineering and Technology*, 32(11), 1831–1844.

Klais, O., Westphal, F., Benaissa, W., Carson, D. (2009b). Guidance on safety/health for process intensification including MS design, part ii: Explosion hazards. *Chemical Engineering and Technology*, 32(12), 1966–1973.

Klais, O., Westphal, F., Benaissa, W., Carson, D., Albrecht, J. (2010a). Guidance on safety/health for process intensification including MS design, part iii: Risk analysis. *Chemical Engineering and Technology*, 33(3), 444–454.

Klais, O., Albrecht, J., Carson, D., Kraut, M., Lob, P., Minnich, C., Olschewski, F., Reimers, C., Simoncelli, A., Verdingen, M. (2010b). Guidance on safety/health for process intensification including MS design, part iv: Case studies. *Chemical Engineering and Technology*, 33(7), 1159–1168.

Klein, T. and Viard, R. (2013). Process safety performances in chemical industry – What makes it a success story and what did we learn so far? In *14th International Symposium on Loss Prevention and Safety Promotion in the the Process Industries*. AIDIC, Florence.

Kletz, T.A. (1991). *Plant Design for Safety: A User-friendly Approach*. Hemisphere Publishing Corporation, New York.

Kletz, T.A. (1998). *Process Plants: A Handbook for Inherently Safer Design*, 2nd edition. Taylor and Francis, Philadelphia.

Krummradt, H., Kopp, U., Stoldt, J. (2000). Experience with the use of microreactors in organic synthesis. In *3rd International Conference on Microreaction Technology*. Springer, Berlin.

Labbe, F. (2006). L'établissement du document unique. *Santé et sécurité au travail*, 6, 6–7.

Laciok, V., Sikorova, K., Fabiano, B., Bernatik, A. (2021). Trends and opportunities of tertiary education in safety engineering moving towards safety 4.0. *Sustainability*, 13(524), 1–22.

Lai, I.K., Hung, S.B., Hung, W.J., Yu, C.C., Lee, M.J., Huang, H.P. (2007). Design and control of reactive distilation for ethyl and isopropyl acetates production with azeotropic feeds. *Chemical Engineering Science*, 62, 878–898.

Langevin, V. and Guyot, S. (2020). *Évaluer les facteurs de risques psychosociaux : l'outil RP DU ED 6403*. INRS, Paris.

Lantz, A. (1998). Fluorinations. *Techniques de l'Ingénieur*, J 5670, 1–10.

Laurent, A. (2011). *Sécurité des procédés chimiques*, 2nd edition. Lavoisier, Paris.

Laurent, A. and Fabiano, B. (2022). A critical perspective on the impact of industry 4.0's new professional safety management skills on process safety education. *Chemical Engineering Transactions*, 91, 1–6.

Lee, J., Bagheri, B., Kao, H.A. (2015). A cyberphysical system for industry 4.0 based manufacturing system. *Manufacturing Letters*, 3, 18–23.

Lepeu, N. (2001). Anhydride phtalique : procédé à moyenne température. *Techniques de l'Ingénieur*, J 6155, 1–15.

Leso, V., Fontana, L., Iavicoli, I. (2018). The occupational health and safety dimension of Industry 4.0. *Medicina del Lavoro*, 109(5), 327–338.

Lins, T. and Oliveira, R.A.R. (2020). Cyber-physical production systems retrofitting in context of industry 4.0. *Computers and Industrial Engineering*, 139, 1–13.

Liu, Z., Xie, K., Li, L., Chen, Y. (2020). A paradigm of safety management in industry 4.0. *Systems Research and Behavioral Science*, 37, 632–645.

Lomel, S., Falk, L., Commenge, J.M., Houzelot, J.L., Ramdani, K. (2006). The microreactor: A systematic and efficient tool for the transition from batch to continuous process? *Chemical Engineering Research and Design*, 84(A5), 363–369.

Lund, M.S., Solhaug, B., Stolen, K. (2010). *Model Driven Risk Analysis: The CORAS Approach*. Springer Science and Business Media, Heidelberg.

Lutze, P., Gani, R., Woodley, J.M. (2010). Process intensification: A perspective on process synthesis. *Chemical Engineering and Processing: Process Intensification*, 49, 547–558.

Lutze, P., Babi, D., Woodley, J.M., Gani, R. (2013). Phenomena based methodology for process synthesis incorporating process intensification. *Industrial and Engineering Chemistry Research*, 52, 7127–7144.

Luyben, W.L. and Hendershot, D.C. (2004). Dynamic disadvantages of intensification in inherently safer process design. *Industrial and Engineering Chemistry Research*, 43(2), 384–396.

Lyu, X., Ding, Y., Yang, S.H. (2019). Safety and security risk assessment in cyber physical systems. *IET Cyber Physical Systems: Theory and Applications*, 4(3), 221–232.

Markert, F., Nivolianitou, Z., Christou, M. (2000). ASSURANCE – A benchmark exercise on risk analysis of chemical installations. In *The 2nd Internet Conference on Process Safety, March 20–24*. Safetynet – European Network on Processes.

Masse, F., Abdo, H., Flaus, J.M. (2018). Comment intégrer les cyberattaques dans l'évaluation globale des risques pour les installations classées ? Proposition d'un cadre général d'analyse des risques. In *Congrès de Maîtrise des risques et de sûreté de fonctionnement*. IMdR, Reims.

Masse, F., Abdo, H., Flaus, J.M. (2019). Une approche de l'analyse des risques des systèmes de contrôle industriels combinant sûreté et sécurité : le cyber nud papillon. *Archives HAL – Ineris*, 02044863, 30–31.

Matlosz, M. (2009). IMPULSE – Multiscale process units with locally structured elements. Technical report, CNRS [Online]. Available at: https://cordis.ec.europa.eu/docs/results/11/011816.

Mazri, C. (2015). Pilotage de la sécurité par les indicateurs de performance. Guide à l'attention des ICPE. Technical report, INERIS, Verneuil-en-Halatte.

Medhammer, I., Bertrais, S., Witt, K. (2021). Psychosocial work exposure and health outcomes: A meta-review of 72 literature reviews with meta-analysis. *Scandinavian Journal of Work, Environment and Health*, 47(7), 489–508.

Michel, L. (2010). Circulaire du 10 mai 2010 récapitulant les règles méthodologiques applicables aux études de dangers, à l'appréciation de la démarche de réduction des risques à la source et aux plans de prévention des risques technologiques (PPRT) dans les installations classées en application de la loi du 30 juillet 2003. *Bulletin Officiel du MEEDDM*.

Moktadir, M.A., Ali, S.M., Kusi-Sarpong, S., Shaikh, M.A.A. (2018). Assessing challenges for implementating industry 4.0: Implications for process safety and environmental protection. *Process Safety and Environmental Protection*, 117, 730–741.

Moustaine, E.E. and Laurent, M. (2012). Systèmes et techniques RFID – Risques et solutions de sécurité. *Techniques de l'Ingénieur*, H 5325, 1–16.

Musu, C., Popescu, V., Giusto, D. (2014). Workplace safety monitoring using RFID sensors. In *22nd Telecommunication Forum*. TELCOR, Belgrade.

Nurliyani, A., Syamsuar, D., Mirza, A.H. (2019). Assessment IT risk management at the computer and network management. *Journal of Informatics and Telecommunication Engineering*, 3(1), 115–124.

OECD (2008). Document d'orientation sur les indicateurs de performance en matière de sécurité pour la prévention, la préparation et l'intervention en matière d'accidents chimiques. OECD, Direction de l'environnement, Paris.

OSHA (2019). Using leading indicators to improve safety and health outcomes. Report, OSHA, Washington.

Pasman, H. and Fabiano, B. (2021). The Delft 1974 and 2019 European Loss Prevention Symposia: Highlights and an impression of process safety evolutionary changes from the 1st to 16th LPS. *Process Safety and Environmental Protection*, 147, 80–91.

Pasman, H. and Knegtering, B. (2013). The safety barometer – How safe is my plant today? Is instantaneously measuring safety level utopia or realizable? *Journal of Loss Prevention in the Process Industries*, 26, 821–829.

Pasman, H., Knegtering, B., Rogers, W. (2013). A holistic approach to control process safety risk: Possible ways forward. *Reliability Engineering and System Safety*, 117, 21–29.

Pasman, H., Rogers, W., Mannan, M. (2017). Risk assessment: What is worth? Shall we just do away with it, or can it do a better job? *Safety Science*, 99, 140–155.

Peruzzini, M., Grandi, F., Pellicciari, M. (2020). Exploring the potential of operator 4.0 interface and monitoring. *Computers and Industrial Engineering*, 139(105600), 1–16.

Petrusich, J. and Schwarz, H.V. (2017). *Industry 4.0 for Process Safety Handbook*. PWL, LLC.

Quintino, A. (2011). What's wrong with risk matrices? Decoding a Louis Anthony Cox paper. In *Decision Support Model Presentation*, Technical University, Lisbon, 1–26.

Ravallec, C., Bondeelle, A., Brasseur, G., Courbon, L., Duval, C. (2013). Risques psychosociaux et document unique d'évaluation des risques. *Travail et Sécurité*, 742, 1–27.

Rogers, A., Evans, R., Wright, M. (2009). Leading indicators for assessing reduction in risk of long latence diseases. Technical report, Berkshire.

Roule, J.C. (2010). *Threat Assessment and Remediation Analysis TARA*. Club SSIF, Paris.

Salvi, O. and Bernuchon, E. (2003). Outils d'analyse des risques générés par une installation industrielle. Technical report, Verneuil-en-Halatte.

Salvi, O. and Debray, B. (2006). A global view on ARAMIS: A risk assessment methodology for industries in the framework of the Seveso 2 directive. *Journal of Hazardous Materials*, 130(3), 187–199.

Schulte, P., Streit, J., Sheriff, F., Delclos, G., Felknor, S., Tamers, S., Fendinger, S., Grosch, J., Sala, R. (2020). Potential scenarios and hazards in the work of the future: A systematic review of the peer reviewed and grey literature. *Annals of Work Exposures and Health*, 64(8), 786–816.

Schuster, A. (2005). Simulation and design for a multiphase microreactor for the direct fluorination of toluene. PhD Thesis, University of Edinburgh.

Schuster, A., Sefiane, K., Ponton, J. (2008). Multiphase mass transport in mini/micro channels microreactor. *Chemical Engineering Research and Design*, 96, 527–534.

Sheehan, B., Murphy, F., Mullins, M., Pyan, C. (2019). Connected and autonomous vehicles: A cyberrisk classification framework. *Transportation Research: Policy and Practice*, 124, 523–536.

Sole, M., Musu, C., Boi, F., Giusto, D., Popescu, V. (2013). RFID sensor network for workplace safety management. In *18th Conference of Emerging Technologies and Factory Automation*. EFTA/IEEE, Cagliari.

Stack, R.J. (2009). Evaluating non independent protection layers. *Process Safety Progress*, 28(4), 317–324.

Stankiewicz, A. and Moulijn, J.A. (2003). *Reengineering the Chemical Process Plant: Process Intensification*. Marcel Dekker, New York.

Sultana, S. and Haugen, S. (2022). Development of an inherent system safety index (ISSI) for ranking of chemical processes at the concept development stage. *Journal of Hazardous Materials*, 421(126590), 1–15.

Swan, R. (2000). Chips with everything. *The Chemical Engineer*, 709, 29–30.

Swuste, P., Theunissen, J., Schmitz, P., Reniers, G., Blokland, P. (2016). Process safety indicators: A review of literature. *Journal of Loss Prevention in the Process Industries*, 40, 162–173.

Tantawy, A., Abdelwahed, S., Erradi, A. (2019). A modified layer of protection analysis for cyber-physical systems security. In *4th International Conference on System Reliability and Safety*. ICSRS, Rome.

Tantawy, A., Abdelwahed, S., Erradi, A. (2020a). Cyber LOPA: A new approach for CPS safety design in the presence of cyber attacks [Online]. Availabe at: arXiv:200600165. [Accessed 6 June 2020].

Tantawy, A., Abdelwahed, S., Erradi, A., Shaban, K. (2020b). Model-based risk assessment for cyberphysical systems security. *Computers and Security*, 101864, 1–15.

Tazi, D. (2014). Les indicateurs de sécurité au travail ne renseignent pas sur la maîtrise des risques majeurs. *Conviction – ICSI*, 4, 1–2.

Thienen, S.V., Clinton, A., Mahto, M., Sniderman, B. (2016). *Industry 4.0 and the Chemical Industry*. Deloitte University Press, New York.

Tihay, D. (2012). Application de la RFID à la prévention des risques professionnels en entreprise. *Hygiène et sécurité du travail*, 2352(226), 11–25.

UICh (1980). Présentation des différentes méthodes d'analyse de sécurité dans la conception d'une installation chimique : l'analyse préliminaire des risques. In *Les Cahiers de Sécurité*, no. 1. Paris.

UICh (2017). Indicateurs de sécurité des procédés. Technical document, ICCA, Paris.

Urban, W., Lukaszewicz, K., Krawczyk-Dembicka, E. (2020). Application of industry 4.0 to the product development process in project-type production. *Energies*, 13(5553), 1–20.

Voynet, D. (2000). Arrêté du 10 mai 2000 relatif à la prévention des accidents majeurs impliquant des substances ou des préparations dangereuses présentes dans certaines catégories d'installations classées pour la protection de l'environnement soumises à autorisation. *Journal Officiel*, 141, 9226–9249.

Wangen, G., Hallstensen, C., Snekkenes, E. (2018). A framework for estimating information security risk assessment. *International Journal of Information Society*, 17, 681–699.

Wen, H., Khan, F., Amin, M.T., Halim, S.Z. (2022). Myths and misconceptions of data-driven methods: Applications to process safety analysis. *Computers and Chemical Engineering*, 258(107639), 1–11.

Wynn, J.E. (2014). Threat Assessment and Remediation Analysis TARA. Technical document. Mitre Corporation, Bedford.

Yazar, Z. (2021). *A Qualitative Risk Analysis and Management Tool CRAMM*. SANS, Paris.

Zwingelstein, G. (2014). Évaluation de la criticité des équipements – Métriques et indicateurs de performance. *Techniques de l'Ingénieur*, SE 4006, 1–27.

Zwingelstein, G. (2019). La maintenance prédictive intelligente pour l'industrie 4.0. *Techniques de l'Ingénieur*, MT 9572, 1–29.

# Index

## A, C, D



## E, I

## K, M, O



## P, R

## S, T, V

Other titles from

in

Chemical Engineering

## 2020

BELAADI Salah
*Thermodynamic Processes 1: Systems without Physical State Change*
*Thermodynamic Processes 2: State and Energy Change Systems*

DAL PONT Jean-Pierre, DEBACQ Marie
*Process Industries 1: Sustainability, Managerial and Scientific Fundamentals*
*Process Industries 2: Digitalization, a New Key Driver for Industrial Management*

SCHAER Éric, ANDRÉ Jean-Claude
*Process Engineering Renewal 1: Background and Training*
*Process Engineering Renewal 2: Research*
*Process Engineering Renewal 3: Prospects*

## 2019

HARMAND Jérôme, LOBRY Claude, RAPAPORT Alain, SARI Tewfik
*Chemostat and Bioprocesses Set*
*Volume 3 – Optimal Control in Bioprocesses: Pontryagin's Maximum Principle in Practice*

## 2018

LOBRY Claude
*Chemostat and Bioprocesses Set*
*Volume 2 – The Consumer–Resource Relationship: Mathematical Modeling*

## 2017

HARMAND Jérôme, LOBRY Claude, RAPAPORT Alain, SARI Tewfik
*Chemostat and Bioprocesses Set*
*Volume 1 – The Chemostat: Mathematical Theory of Microorganims Cultures*

## 2016

SOUSTELLE Michel
*Chemical Thermodynamics Set*
*Volume 5 – Phase Transformations*
*Volume 6 – Ionic and Electrochemical Equilibria*
*Volume 7 – Thermodynamics of Surfaces and Capillary Systems*

## 2015

SOUSTELLE Michel
*Chemical Thermodynamics Set*
*Volume 1 – Phase Modeling Tools*
*Volume 2 – Modeling of Liquid Phases*
*Volume 3 – Thermodynamic Modeling of Solid Phases*
*Volume 4 – Chemical Equilibria*

## 2014

DAL PONT Jean-Pierre, AZZARO-PANTEL Catherine
*New Approaches to the Process Industries: The Manufacturing Plant of the Future*

HAMAIDE Thierry, DETERRE Rémi, FELLER Jean-François
*Environmental Impact of Polymers*

## 2012

DAL PONT Jean-Pierre
*Process Engineering and Industrial Management*

## 2011

SOUSTELLE Michel
*An Introduction to Chemical Kinetics*

## 2010

SOUSTELLE Michel
*Handbook of Heterogenous Kinetics*